Trefethen's index cards
Forty years of notes about
People, Words and Mathematics

Trefethen's index cards
Forty years of notes about People, Words and Mathematics

Lloyd N. Trefethen
Mathematical Institute, University of Oxford

World Scientific

NEW JERSEY · LONDON · SINGAPORE · BEIJING · SHANGHAI · HONG KONG · TAIPEI · CHENNAI

Published by

World Scientific Publishing Co. Pte. Ltd.
5 Toh Tuck Link, Singapore 596224
USA office: 27 Warren Street, Suite 401-402, Hackensack, NJ 07601
UK office: 57 Shelton Street, Covent Garden, London WC2H 9HE

TREFETHEN'S INDEX CARDS
Forty Years of Notes about People, Words and Mathematics

Copyright © 2011 by World Scientific Publishing Co. Pte. Ltd.

All rights reserved. This book, or parts thereof, may not be reproduced in any form or by any means, electronic or mechanical, including photocopying, recording or any information storage and retrieval system now known or to be invented, without written permission from the Publisher.

For photocopying of material in this volume, please pay a copying fee through the Copyright Clearance Center, Inc., 222 Rosewood Drive, Danvers, MA 01923, USA. In this case permission to photocopy is not required from the publisher.

ISBN-13 978-981-4360-69-2 (pbk)
ISBN-10 981-4360-69-4 (pbk)

Printed by FuIsland Offset Printing (S) Pte Ltd. Singapore

For Emma and Jacob, who understand

Foreword

Every so often a great scientist or mathematician lets us in. All the way. Not just about his work, but about everything he ponders and cares about.

He reveals who he is at heart.

Jim Watson did it in *The Double Helix*, and so did Stan Ulam in *Adventures of a Mathematician*. As a youngster reading those books I remember being shocked by both of them – for their iconoclasm, for their unapologetic brilliance, for their insider's view of science at the top, and above all, for their idiosyncratic and often startling observations about life. The effect of inhabiting a mind like this was bracing.

Now Nick Trefethen joins their ranks with this remarkable new book. It's wry, intimate, and at times, disconcerting. Nick's true self is on every page, whether wistfully describing his father's book collection as a metaphor for how alone we all are, or in ruminating about fatherhood, sex, dog toilets, and the magic of the number ten billion.

What's especially original here is the book's structure. It's a collection of thoughts and questions, some playful, some very deep, each compact enough to fit on an index card. Nick has been writing these index cards to himself for the past 40 years. By arranging them longitudinally, he allows us to watch him unfold, captured as if by time-lapse photography, as he matures from promising teenager to the Professor of Numerical Analysis and FRS at Oxford.

Whether you're a fellow mathematician, or merely a fellow human being, you're in for a treat you'll never forget. I know of nothing else like it.

Steven Strogatz
Cornell University

Preface

This is a book about ideas, and also about one mathematician's personal development. I hope you find the mix interesting.

I was born on August 30, 1955 and started writing notes on index cards in February 1970. I remember that day, with my typewriter and a pack of 3"× 5" cards, thinking what a good thing it would be to record some thoughts that kept recurring in my mind. I was 14, a fruitful age for young philosophers, and the pack grew quickly! Later I switched to 4"× 6".

Forty years on, I'm still writing notes, having settled down to a rate of two or three a month. The basic aim hasn't changed: nothing less than to collect little nuggets of truth, to play the Glass Bead Game. Of course, as the years go by, one acquires a more realistic view of how much truth one man can muster. And I've taken growing pleasure in the sideshows — the snapshots of history, the tidy turns of phrase, the quirks of personality.

Why doesn't everybody feel the need to organize their thoughts in a card file? Half of me remains puzzled about this while the other half knows the answer pretty well.

For me, at any rate, this medium is the right one. Once I've put an idea on a card, it becomes a piece of my mental framework, a principle I will refer to for the rest of my life.

To make sense of these notes it may help to know that I was married to Anne Trefethen from 1988 to 2007 and that our children are Emma, born in 1991, and Jacob, born in 1993. I was in grade school to 1970 (Shady Hill), high school to 1973 (Phillips Exeter), university to 1977 (Harvard), graduate school to 1982 (Stanford), and then employed at NYU (1982-84), MIT (1984-91), Cornell (1991-97), and Oxford (1997-present). Thus I lived in America until 1997 and in England since then. I belong to both countries and love them both (see p. 99). Interruptions and sabbaticals took me to Australia and elsewhere in 1971-72, 1976, 1979, 1982, 1990-91, and 2003-04. By profession I am an academic, an applied mathematician. More precisely I am a numerical analyst, which means I work on algorithms for solving mathematical problems on computers.

Some of the clumsy wordings in these notes bother me, especially in the early years, as do some "he"s and "him"s. But in preparing the selection for this book, I have changed little apart from anonymizing some names, removing cross-references, and shortening here and there. The section titles and organization are new, but the original titles of the notes themselves, those that had them, have mostly been retained. On the theory that adulthood begins at 20, I've listed my age for those notes written as a teenager, and where possible, I start each section with a note from age 14 or 15. Within each section, the ordering is chronological.

About the picture on the cover, see p. 198.

Nick Trefethen
Oxford, April 2011

Contents

Foreword by Steven Strogatz

Preface

Ego *1*

Kids *11*

Aging and Death *21*

Sex *33*

Living with Others *43*

The Meaning of Life *55*

Politics and Society *63*

Cold War Nukes *73*

Education *83*

Britain *93*

Famous People *103*

Optimizing Your Life *115*

The Life of the Professor *129*

Music *137*

Words *145*

Writing and Literature *157*

Memory *169*

Misperceptions *177*

Knowledge and Truth *187*

Analogies *201*

Bad Logic *213*

God and Religion *225*

Good and Evil *239*

Science *251*

Stars and Planets *261*

Mathematics *271*

Big Numbers *289*

Mathematics and Science in Everyday Life *299*

Inventions *313*

Computers *323*

Life and DNA *333*

Hearts Minds and Bodies *347*

Index *359*

14 February 1970

Index card #1

I cannot bear the idea of going through life without being the most effective person who ever lived. Then my life will be satisfactory, otherwise it will be wasted.

(Age 14)

29 December 1982

What's the point of striving?

Churchill (I think) once said, more or less, "Democracy is the worst form of government, except for all of the alternatives."

This is the sort of argument by which I justify my personal ambition, the fact that I organize my life about a narrow and academic set of goals. I am not Mozart or Einstein; the historical significance of the things I accomplish through all this hard work will probably be small. Thus in a sense it is absurd that I take myself and my career so seriously. Yet what would you have me replace this modest but well-organized life with? A random sequence of pleasurable actions?

12 November 1986

Goals for August 30, 1990

Here are four ambitions that I hope to achieve by my 35th birthday:

1. I will be married to a wonderful woman.
2. I will be tenured at an outstanding university.
3. I will speak German and French comfortably.
4. I will be an accomplished piano player and skillful sight-reader.

24 August 1998

Truth and ego

Scientific civilizations, I have become convinced, do not last very long. I no longer believe in a distant future for the human race, and the sense of darkness ahead hangs over all my thoughts.

You couldn't ask for a larger subject, but do you know something? Through it all, my ego is so swollen that I am caught up not merely in the power of this depressing idea, but in a desire to get credit for it. Maybe I'm not the first to discern that doom is probable, but I'm one of a small group, and nobody has yet argued the case properly to the public. Maybe I can be the one! I must publish! Put my name on this idea!

As the calamity strikes that finally extinguishes our civilization, the last sound heard will be Trefethen shouting, "I told you so!"

4 December 2001

My father's book collection

As my parents moved out of the house they'd lived in so many years, we sold and gave away most of their books. This process was painful for my father. To him his books were not just the sum of their titles — they were a collection, a lifetime's orderly organizing of science fiction and Victorian fluid mechanics and other special topics. How could it be that the book buyers didn't appreciate this? To them the books were just individual titles.

My father's book collection symbolizes to me how alone we are in this world. Yes, I have good conversations with my wife and my children and my friends. But in the end they don't follow the full pattern of my thinking, just some pieces. On the inside, an active mind feels like a coherent pattern. An outsider sees it only partly, and when the life is gone, the pattern is lost forever.

4 January 2003

The fleeting exhilaration of big honors

Fred Hoyle in his autobiography suggests that honors like election to the Royal Society have mainly a negative effect on people. He contrasts his tension in the years before the good news finally came when he was 41 to his lack of feeling afterwards. "Oddly enough, by the end of that week, the exhilaration I had felt initially was gone, and since then, I have never really thought anything much of it."

Yes indeed, but isn't he omitting a part of the calculation? What we get from honors is not just fleeting exhilaration, but a lasting validation that our great efforts have some meaning, that we are the special people we strive to be. For people of ambition, honors contribute to the quiet confidence that is an important part of being happy and successful.

You never notice the value of your ankle till after you've twisted it. Does this mean it isn't valuable?

1 July 2006

World Cup football and life

This evening Jacob and I watched two agonizing World Cup quarter-finals. Portugal beat England by penalties; if only Gerrard had aimed at the other end of the net, the result might have been different. Then France and Brazil played a marvelous game, but Ronaldinho's free kick rose a heartbreaking foot above the goal bar, whereas when Zidane made his perfect pass, Thierry Henry was there. So Brazil is out.

Life for a World Cup footballer is like life for the rest of us, but concentrated. How does it feel to be one of these players, a thousand times richer and more famous than the rest of us, subject to the same career ups and downs yet a thousand times faster? Just as for you and me, skill and ambition explain the big gaps (Brazil-Paraguay) but the narrower ones have as much to do with chance (Brazil-France). At first one is struck by the awful fact that every team but one must lose. But isn't that, more slowly, the same for you and me? I'm not the world's #1 mathematician, and Ashley Cole isn't the world's #1 footballer, but I think we're both pretty well satisfied with our careers.

Yet there is a difference between soccer and life. In soccer, the near-misses are agonizingly apparent, and the worst ones get replayed over and over on television. In life, we notice a few of the near-misses but are unaware of most of them.

11 October 2008

The Copernican Principle and my job

Employers and employable skills change so fast in this modern era, nobody can feel secure. What if my company folds? What if my specialty becomes obsolete? I write at a time of particular economic crisis.

According to Gott's Copernican Principle, if you want to estimate how long something will last, a pretty good starting point is to suppose that it will last as long into the future as it has lasted already. The principle predicts that the USA will probably survive for centuries to come, for example, whereas Bosnia and Herzegovina might only be around for a couple of decades.

Well I guess I'm reasonably safe. My employer, Oxford University, is 750 years old. My profession, mathematics, was already strong by 250 BC.

30 May 2009

Fame and pinging

Steve Strogatz came up in conversation. Steve has become known for his exceptionally clear communication of mathematical ideas. How's this for the compliment I heard about his book *Nonlinear Dynamics and Chaos*? I sent him an email, knowing he'd like it.

> "It's the best book on any subject I've ever read!"

But fame is a funny thing. We yearn to be influential, and it means so much to us that people have been impressed or amused by our creations. We want to be aware of their happy reactions, and we love a sincere compliment. Yet the last thing Steve and I would want is to be pinged *every* time somebody enjoyed one of our paragraphs. The optimal rate of feedback is not 0%, but it is closer to that than to 100%.

22 August 1995

Why are sleeping babies so hard to wake?

A sleeping baby can be hilariously hard to wake up. Hand claps, thunder claps, heavy artillery — is there anything they won't sleep through?

Here's a question. Do babies have this property, evolutionarily speaking, because they are helpless? Because a creature that can't defend itself has little to gain from waking up in the presence of danger?

And here's a meta-question. Will it soon become easy to find out the status of a question like that? At present, if I were determined to know whether this theory has been proposed before and whether the facts support it, I would look around in the library and start asking biologists and psychologists which of their colleagues I should speak to. It would take many hours and many conversations to find the state of the art, with no guarantee of success. Consequently, since I am not determined but merely curious, I won't do it.

In the future, through networked computing, will all this change? At stake is not just babies' sleeping behavior but, for starters, half of these index cards accumulated over a period of 25 years.

10 October 1996

Adults' and children's visceral feelings

Many a movie and many a news story has a parent who loses a young child. If you're a parent yourself, nothing could be more wrenching. Scenes of this kind fill Anne and me with such strong emotions that we look each other in the eyes, generally damp, sharing the feeling and the wonder of its power. If Emma, 5, is watching too, she doesn't understand how deeply we are affected. "When you're a mother," we explain, "you'll understand."

Something odd happened the other night as we watched a movie with Emma. It was a lightweight kind of drama, easy to laugh at despite a full complement of lions and snakes and poisonous spiders ("Jumanji"). Suddenly, at an ordinary enough moment of the plot, we became aware that Emma was having a big problem. She was deeply upset, on the verge of tears, clutching us for protection and comfort. With awe at the workings of emotions, we recognized the reason: in this scene, a young child was losing its parents.

22 September 1997

Children and sailors

Anne and I got talking about sailors with Emma (6) and Jacob (4). I found myself explaining that traditionally, sailors on ships are made to work very hard. Work, work, work is the policy, and the reason is, if you don't keep sailors busy, they start causing trouble. They get loud, they do crazy things, they start fighting. Trouble! Best to keep them busy all day, so they fall straight into a deep sleep at night.

All this, the four of us realized more or less at once, sounds awfully familiar. Ask any parent.

2 June 1998

France and England, men and women

Emma asked as we explored the villages above St. Etienne, why do you and Mummy keep going on about how different France is from England? It doesn't look that different to me.

Well, I answered, of course, you're right in a way, but it's like the difference between men and women. To a nonhuman observer the two sexes would appear much the same, differing only in minor details. But how powerfully different they seem to us humans!

15 August 2000

A day in the life of Emma at Oxford

Emma spent an hour at my office. "Daddy, may I try your shredder?"

She did, and kept the quarter-inch paper strips that resulted. "I could have used some old ones from the basket, but I thought it would be more fun to make my own."

It turned out her aim was weaving. Soon she had a woven square, flexible and strong.

We left, and I asked her to carry Jacob's cello to the car as my hands were full. "I'll use this as a Jacob protector, so I don't have to touch his icky cello with my hands."

We went to the Nosebag for lunch. We joked about the Jacob protector and talked about how cutting on the bias makes garments cling. Our own shirts didn't cling.

I noticed Emma's woven square was 9x10, and that she'd marked X's checkerboard style. Emma (hiding it under the table), how many X's are there all together?

"46!" She grinned. And showed me that she'd made a mistake in her X-ing, put a cross in one of the squares that should have been blank.

She finished her chocolate cake. We went to the Ashmolean and looked at Turner's Oxford and the blank eyes of Roman sculptures.

18 October 2000

Safe biking with Jacob

7-year-old Jacob was upset that I made him walk his bicycle on the sidewalk to avoid a dangerous bit of traffic on St. Clements. "I'm as safe a biker as you!", he complained.

This set me thinking, and I realized there are at least four reasons why I apply stricter rules to his biking behavior than to mine.

First, years of experience have given me better instincts than his of how to handle a bicycle.

Second, years of experience have given me better perception of traffic hazards.

Third, I have the conservative movements of an adult, while Jacob still attracts random accidents like any child.

Finally, there's no denying it. I value his life more than my own. If Jacob heard this last argument he'd complain, "That's not fair!"

5 April 2001

Chewing on guns, germs, and steel

Jared Diamond's *Guns, Germs, and Steel* is one of those thick books that would look untouchable to a child but is gripping reading for an adult like me. I have my doubts about some of his answers, but what great questions!

Early in the book Diamond tells the story that among certain tribes in New Guinea, the old people generally have no teeth left. So the youngsters chew food and give it to their parents and grandparents, pre-chewed.

I told the children this cross-generational story, and you bet they were interested.

Only later did it occur to me that of course they didn't read the story themselves, complete with all the words and context of Diamond's larger purpose. No, I gave it to them pre-chewed.

10 April 2003

How to bring back your childhood

The other day Anne was driving and I was in the passenger seat. As an experiment I tried something extraordinarily unusual. Instead of looking forward at the road, I turned my head 90° and looked resolutely out the side window. Not just for a second or two, but for the best part of a minute. It felt dangerous, and irresponsible. The roadside rushed by.

My childhood came back to me. I had a powerful sensation of being six or seven years old, driving along on a family trip through Maine or New England with Gwyned and my parents.

26 August 2006

Jacob and my beard

In 1999 we had a two-week holiday in Ireland. As usual on holidays, I didn't shave. At the end I had a nice fuzzy beard, but I shaved it off on our final day. Six-year-old Jacob was brokenhearted. He cried and cried.

Now Jacob is 13, and we're coming to the end of a two-week holiday in Iceland. Should I shave off my beard, I asked?

"It's your face," he said.

Aging and Death

14 February 1970

The horror of middle age

I have long had a monstrous fear of death and especially aging. It is the thought that everybody and particularly myself has only one life, and only one chance to live it. I dread getting a year older thinking I could have accomplished more, and the thought of middle age, before one has begun to deteriorate physically, but yet has finished one's beginning of life, fills me with unspeakable dread and horror.

(Age 14)

23 February 1988

The chief preoccupation of adult life

The chief preoccupation of adult life is the replenishment of the appetites.

7 February 1998

Writing and aging

Many of us build our lives around writing, spending our hours, year after year, turning thoughts into words and polishing them till they shine.

As the years go by and memory becomes less reliable, I think the habit of writing takes on a special significance. In interacting with the written page, we can edit and adjust and keep on track even at an age when on the hoof, our thoughts would ramble and we'd be at a loss to recall every third name. Our written words are a comfort, a flywheel that enables our ideas still to move forward even as the engine declines in power.

We saw an extreme example on New Year's Eve with P., now 84 years old. "Jessie," he asked his wife three times in the course of the evening, "what's the name of that chap I'm writing a book about?"

1 May 2000

Why women live longer than men

Walking around Five Fields the other day, 9-year-old Emma and I were discussing the fact that women, on average, live much longer than men. I explained that so far as I knew, though pieces of an explanation were in place, the causes of this difference were not fully understood.

"Maybe," she suggested, "it's to make up for all the other stuff."

5 December 2001

Life after death

My father died last month, but you don't absorb a thing like that instantly. If you live far away, you may have to force yourself to remember that he isn't at the end of a phone any more. It's as if his death isn't just a fact handed me on a plate, but something I have to practice and learn till I've got it by heart.

This experience has changed my conception of why people think there is life after death. Previously I had imagined that this idea was a pure fabrication, a big bite of irrationality whose emotional origin was clear enough but whose epistemology was laughable. Now I see that mere carelessness and inertia, epistemological errors of omission rather than commission, are enough to get you half way there.

3 March 2002

The diversity of old age

Though children and adults are not all the same, most of the variation lies along a few familiar dimensions. But what diversity we experience in our declining years! The reason is obvious. Fate snatches away our capabilities in arbitrary order, and the lives it generates in the process, like Tolstoy's unhappy families, are strongly drawn. One old-timer is half-blind, another half-deaf; one stuck in a wheelchair, another fuzzy in the head. What binds them together is the warm consciousness that we all must ride down the cobbled path.

15 July 2003

The dual of age

We all have two ages, equally interesting: the time since our birth, and the time until our death. We talk only about the first, because the second is unknown. There's not even a word for it, so let's call it our *remainder*.

Imagine how different a room full of people would look if we knew everybody's remainder as well as their age! Jim here is young; he's got thirty years ahead of him. Joe, alas, is equally fit, but as you can see, he'll be gone the year after next.

22 August 2004

Graceful aging

You can't be young forever — no 60-year-old can act 20 without looking silly. The proper aim is relative: at each stage of life, to be a good example of that stage. You can't be a kid, but be young for your age.

27 December 2004

Carpe diem is just the first step

The classic idea is that life is fleeting and youth even more so, so we'd better grab our pleasures while we can. "Gather ye rosebuds while ye may, old time is still a-flying."

I think the actual situation is a good deal worse. I've gathered plenty of rosebuds in my day, but mostly I can't *remember* them! What's the point of all life's adventures when 99% are soon buried and forgotten? You can palliate the effect by keeping diaries and photo albums and having a spouse to reminisce with, but this just improves the score to 98%.

Gather ye rosebuds and smell them deep, the past is awfully fuzzy.

29 September 2005

Aging and accidents

Old people have a lot of accidents on stairs, in baths, on sidewalks. It seems obvious that this happens because they are physically limited, but I think it's not so simple. I think the problem with older people is that they are limited not in an absolute sense, but relative to the habits of a lifetime. From infancy, we calibrate how fast we can walk or run or trip down the stairs without hurting ourselves: if there's a misstep, we must be fast enough to recover. When the perceptions and muscles slow down in later years, the calibration may no longer match the equipment. The trick to safety in old age is to slow down your timing as fast as your nerves and muscles are slowing down.

11 October 2007

How full is my bowl of ice cream?

There's a misperception I've noticed lately. When I eat a dessert, I feel good almost all the way through knowing it isn't done yet. When I'm 10% finished, it's good knowing there's plenty more to eat. And when I'm 90% finished it feels the same! So long as there's more than one spoonful left, I feel I'm in the middle, doing fine, no problem. Only at the final spoonful am I suddenly grabbed by the pang of running out.

This is a pure and simple illusion, a bug in my thinking that I can observe with interest just as I might observe an optical illusion in my vision. I try to compensate by telling myself, hey, your ice cream is running out, you should be upset about this: but it's no use, my heart keeps feeling plenty though the bowl is nearly clean.

And what about one's course through life itself? I felt my life was full and good at age 8, and I feel that way now, too. The calendar tells me I'm well into my second half, but no problem, there's plenty ahead.

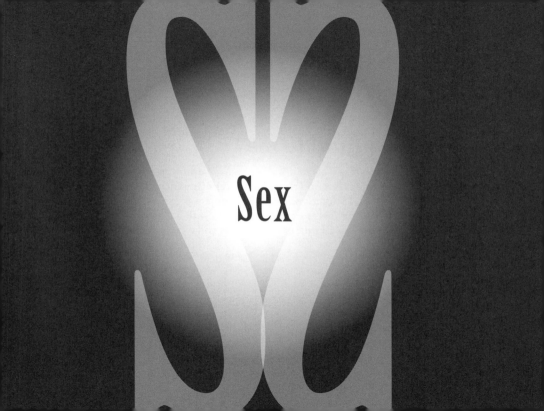

30 November 1970

A girl is beautiful to me

Isn't it remarkable the effect a certain shape can have upon a person? Think of how strange it is that a man does not find excitement in another man, but if a few lines are changed and the skin made slightly softer, it is an object of huge desire. It is strange the extent to which instinct forces us to be obsessed with a form that certainly has no intrinsic beauty. I can say all this, of course, and still a girl is beautiful to me.

(Age 15)

22 February 1985

Cross-country skiing and sex

Cross-country skiing is to downhill as cuddling is to sex. I wouldn't want to live without either, but let's face it, when it comes to intensity of pleasure, there is no contest.

But if you *had* to do without one or the other, which should it be? Reluctantly, I would sacrifice the sex and the cross-country. This sounds unparallel. The difference is, you can downhill ski all day long.

3 June 1985

Male and female sex drive

I believe that men have an innately stronger sex drive than women. This fits my personal experience, but so do many other gender differences that I doubt are innate. What makes this case special is that there is a compelling evolutionary reason why the male sex drive *ought* to be stronger. A woman can have at most about one baby per year, so it is in her interest to emphasize quality rather than quantity in picking mates. A man, by contrast, can have dozens, and it is in his interest to do so.

Evolutionary arguments are advanced to establish all sorts of innate differences between the sexes — usually to show that men are superior. For example, some people argue that men must be more intelligent, since hunting demands more cleverness than tending babies. I think such arguments are usually worthless. Whether it is hunting or tending babies that demands more intelligence is a question far too delicate for us to pronounce upon. An evolutionary argument about gender differences is compelling only if it is founded squarely on the *definition* of the difference between male and female — namely that it is the latter sex in whose body the fetus develops.

24 November 1986

To a loved one in Europe

When do I love thee? Let me count the days.

23 May 1994

Looking at babies and looking at women

Most women, and most fathers, can't keep their eyes off a baby. I know I can't. I beam at babies sitting across from me on the bus and turn my head to peer into baby carriages as they roll by. Chatting to these irresistible creatures is also standard practice. If I hold one-year-old Jacob's hand as he walks down the aisle of the airplane, half the passengers stop him for a gooey interview en route.

Now babies are not the only irresistible creatures we encounter in our daily rounds. In addition, there are women. But what a difference! When you come up against a good-looking woman, all is circumspection. A casual glance or two, that's the limit. No beaming, no instant conversations, and certainly no patting and cuddling.

29 August 2000

Food and sex drives

Our drive for food is zero after a good meal and rises decisively toward infinity in the hours and days afterwards.

Imagine if sex were like that.

21 January 2003

Cats playing with string

If you wiggle a piece of string in front of our cats Tiger and Jack, they are fascinated and can't resist pouncing on it. Does that mean that Tiggy and Jackers are unaware that this isn't a living thing? I very much doubt it. After all, I'm as distracted as the next guy by a page 3 photo or a mannequin with her blouse off. I think I know how the cats feel, enjoying while knowing.

But tell me, why aren't pictures of food interesting in the way that pictures of women are interesting?

2 June 2008

Bodies attract

A paradox of the dynamics of solar systems and galaxies is that all the bodies attract gravitationally, yet they don't crash together. Instead they fly around each other. All that gravitational energy — what it generates is not agglomeration but motion. Depending on the masses and initial conditions, you may see stately orbits or a chaotic dance. Stars may move closer together sometimes, whereupon their speeds increase, but they hardly ever hit.

Talking with 17-year-old Emma, I have realized that a similar rule applies in teenage social dynamics. The girls take good care to be attractive to the boys. You might imagine that their aim must be to pull in a partner, but it's not that simple. The last thing most of these young ladies want is to hook up with somebody permanently. Their beauty is doing its job if it pulls the boys closer, sends them into orbit, keeps the game exciting.

31 January 2010

At the Gauguin exhibit

Every boy at the museum with his mother is struck by all those naked ladies. Why are there so many naked ladies?

If your mother is like mine, she explains that artists are concerned with beauty and the contours of the human body represent one of the highest forms of beauty. It's all quite noble and ethereal.

Later in life you begin to make out some of the other pieces of the explanation. Artists really like women's bodies. So do art buyers, and they are quick to shell out money for nudity legitimized as art. And if you're an artist and have to spend hours with a model, well, there are some clear advantages if she has to be naked.

Living with Others

14 May 1971

One thing you can count on

Much human misery comes from uncertainty. Does she love me? Do they like me? Will I flunk my physics test? Will my life be a failure? And many of these worries stem from uncertainty about other people's feelings.

Amid all this, there is one fact about other people that we *can* take for granted, that is given, that stands out like a bright and happy light against a background of uncertainty: people want to be liked. No matter how much Mr. Thomas hates you, Mr. Thomas hopes you like him. That is nice to know.

(Age 15)

7 February 1972

Lifelines nexus

It is interesting to think, at a gathering of ten or a thousand people, of what a rare and complicated set of events has led to each of these people being present at this time. A dinner party or a ball game is a sort of nexus, and it is tantalizing to imagine the life-lines of these people, together for an instant here-and-now, but diverging ever more widely as you look into the past or the future.

(Age 16)

25 February 1978

The Palo Alto bike trapper strikes again

Each time I park my bike I unhitch my impregnable Citadel lock and secure it through the spokes to keep the bike from being stolen. This is a sensible thing to do.

Well, suppose I went out and fastened my impregnable lock through somebody else's spokes while he wasn't around? Suppose I went on a binge, bought five thousand locks and immobilized the entire Stanford community as it sat in class some morning? The dread Palo Alto bike trapper strikes again!

These would be pure, atomic acts of madness: ordinary in the physical description, but devoid of motivation. I'll never perform them, and the chances are you never thought of them. The natural, well motivated function of things in the world so dominates their use that we come to think of bike locks as inherently defensive devices. But you won't find that defensive nature anywhere in their steel bars and moving tumblers.

4 June 1982

Tourism and tolerance

At a brief exposure the foreign culture may seem more foreign than it is. One naturally fixes on the strange smells and unfamiliar customs, and, particularly if the language is foreign, it may be hard to see that behind it all these people are not much different from us. So a week in Bombay is not guaranteed to make you more of an internationalist; it might make you more of a bigot. Which one it is probably depends largely on the theories in your head with which you interpret the experience.

If you sit at home instead and read a book about India, you miss the immediacy of the confrontation with dark men in turbans, smells of curries, and inscrutable Hindi. As a result, the impression you get may be more realistic. Described in English, the Indians begin to seem more or less like ourselves — which to first approximation is surely the correct conclusion.

19 September 1984

The hostility of a new environment

Whenever I come to live in a new place, my first few days are characterized by feelings of apprehension, even by the sense that this new world is hostile to me. The little things seem in conspiracy to make my life difficult. I have been feeling this sensation during these first few weeks at M.I.T.

As a child I experienced agony of this kind when I was sent to camp in Australia — it's called homesickness. In adult years the worst such experience was during my first few days in Zurich in 1979. The supermarket seemed especially inimical to me, full of mustard in tubes instead of bottles and other malevolent oddities.

The irony is that one's perception that the new environment cares enough to thwart one's interests is precisely backwards. In fact, the new environment is indifferent to you, while the old one was actively benign. After a time in any place, we find hundreds of little tricks, shortcuts, to make life go smoothly. What we perceive as the normal, sensible world is in fact an edited copy from which we have removed most of the troublesome elements.

23 January 1987

Lovers and strangers

Damn it all, why do we snarl
at our partners of opposite sex?
With strangers we add up their virtues;
With lovers we count their defects.

11 April 1987

What do my friends think of her?

As you and your girlfriend become closer, you begin to hear less often what your friends really think of her. Once you announce the engagement, that's pretty much the end of all honest communication on the subject.

I don't like this effect, but I've thought of a way to interpret it that provides some consolation. In love or marriage, two people to a certain extent fuse into one. Your friends stop talking freely about your fiancee in rough proportion as she becomes a part of you. So their reticence can be viewed as a special case of a different phenomenon — long familiar, and probably a good thing for all of us — that people rarely reveal what they really think of *you*.

14 October 2001

100-50 or 50-100?

I've never had a conversation with Nick Trefethen, but I'm guessing I might not like him so much. I have ideas, and I try to push them forward. I'm aggressive about making sure that the other guy has heard and understood. Just to be sure, maybe I'll repeat the point that extra time.

Now I ask myself, hypothetically, consider these two approaches to conversation. You could be assertive, and get your point across 100% of the time while being liked 50%. Or you could hold back and be gentle, get your point across 50% of the time, and be liked 100%.

Which is the right strategy?

26 February 2006

Chutzpah in our back garden

We've all heard that famous example of *chutzpah* — the youth who murders his parents and then begs the judge for leniency on the grounds that he is an orphan.

It turns out such things really happen. When we moved into our house here in Oxford, it looked out back across our neighbors' big green garden. Then they divided the property in two and sold the garden to a developer, who put up a house in it. Now we've applied to the City Council for permission to build a new third-floor guest room. You guessed it. The occupants of the new house have lodged an objection on the grounds that we must "protect this conservation area."

27 September 2008

Too cold? Too hot?

The train carriage Thursday was too hot, and I mentioned this to the attendant. Well, that's a new one, he said. People don't usually complain that it's too *hot!*

We have here a mystery. If the temperature is on average about what people want, then you'd think there should be equal numbers of "too cold" and "too hot" complaints. Yet I suspect the attendant was right and he doesn't hear equal numbers. In fact I am frequently exasperated by hot trains and restaurants and shops and lecture rooms, and perplexed that people put up with them so meekly.

What's going on? I have two theories. Here's the first. Maybe people consider it more aggrieving to be too cold than too hot because at bottom we connect heating with age-old survival needs, with fighting the elements, whereas cooling is a luxury. In other words, though it's nice to be cool, we feel we have a *right* to be warm.

The other theory is that warming up when we are cold is easy (just put on a sweater) whereas cooling off when we are hot is not so easy — so we habitually assume coldness is a problem to be fixed, whereas heat is an inconvenience to be put up with it.

What do you think? Is one of these theories on target?

13 May 2009

Living with a geek

Kate complained, you're always so fast and logical!
OK, I said, I'll try to be slow or illogical.

The Meaning of Life

30 December 1985

Intelligence and survival

The success of advanced life forms depends upon powerful drives to survive and procreate. Natural selection has provided these drives in all the advanced species on earth.

In conflict with such instincts is the fact that survival is pointless. Below a certain level of intelligence, a species is unaware of this fact, so it has no impact. But it seems likely to me that any species advanced enough to maintain a scientific culture must eventually recognize that life has no meaning.

How can it then continue to survive? Somehow the deadly truth must be kept sealed in the intellect, out of reach of the instincts. But I doubt whether such a separation can persist in the long run. Can it be that higher levels of intelligence than ours are possible biologically, yet will never be achieved by natural selection?

1 June 1986

Suicide in the future

Has the meaninglessness of life had an impact on demography? Of course one must note that at present, even educated people still generally believe that life has a meaning. But this conviction is probably eroding as the centuries of science unfold.

The drive to *procreate* has begun to weaken. This is the old complaint of eugenicists — the lower classes outbreed the upper ones. Rich and educated Westerners have few children, and I suspect that at bottom this is a rational decision. Raising children is an enormous effort; we do it anyway because of evolutionarily instilled drives related to sex (increasingly inoperative because of contraception) and nurturing.

The drive to *survive* remains relatively intact. Nobody I know has stopped eating because of the meaninglessness of it. A more surprising observation is that the incidence of suicide in our world is still very small. In a technological society, it becomes inevitable that painless death will eventually become widely available. Couple this with a spreading opinion that there is no categorical objection to it, and it seems to me, one must expect a society in which the leading cause of death is suicide. I am afraid it is likely that this will indeed be the situation within a century or two.

10 September 1986

Is there anything worth talking about?

Like many people with intellectual pretensions, I chafe at the banality of daily conversations. The central topic of human interest seems to be, *how can I live well?* This is the basis of our endless chatter about restaurants, movies, condo prices, even sex.

What might we talk about instead? The alternative I yearn for is, *what is the truth?* The subject should be the universe, not just our narrow culture; the goal should be explanations, not just facts!

Yet I am increasingly cynical about the sharpness of these distinctions. To those who believe in God, or ethics, or the soul, topics for discussion may seem to exist that are fundamentally deeper than condo prices. But I reject such absolutes. Ultimately there is nothing more profound than, say, the laws of physics or biology — and between quantum mechanics and condo prices, there lies a damned continuum. The trick is to maintain one's enthusiasm despite the fact that it's all actually meaningless.

1 December 1986

The meaning of life

Life is pointless, and any fool knows it;
But I've no time for the fool who shows it.

28 June 1997

The meaningfulness of meaninglessness

Our deepest emotions, and our greatest masterpieces of literature, derive their power from the meaninglessness of our existence. The purposelessness of it all, the emptiness of our personal struggles and triumphs — this is the ineffable theme that has moved minds through the centuries. Though believers will deny it, I think this is the wellspring of much of religious experience, too.

If life had a meaning, matters would be different. Our thoughts would be less subtle, less endlessly variegated. There would be one great novel to write, not a thousand.

If God existed, there would be far less to say about Him.

I think of it mathematically. Let $f(x,y)$ be a random signal in the variables x and y. What can you say about the function $1+f(x,y)$? Not much, for it's just a constant plus some noise. But now, what about the function $f(x,y)$ itself? Endless structure! Infinitely fascinating! However closely you examine it, there's always more to learn!

1 March 1999

Pretending there's a meaning (or not)

Many an intellectual, like me, reaches the conclusion after careful thought that life has no meaning. Meanwhile our actions say just the opposite, as we single-mindedly pursue that higher purpose we've selected for ourselves that we are convinced, at bottom, will save our souls.

More ordinary people often believe that life does have a purpose. Meanwhile their actions say just the opposite, as they casually live the life we are too self-conscious to achieve for ourselves, making the best of the day-to-day.

17 November 2003

Hard work and immortality

Why do people like me work so hard? It's an old idea that we're trying to achieve some kind of immortality.

I think there's truth in this, but the immortality has two aspects. The obvious one is that we hope to be remembered after our death. Yes, I'd like to be remembered, but I think there's another aim that is more important. Through my work, I imagine that I transcend the meaningless animal nature of my existence. We strive not just to extend our mortal lives, but to rise above them.

I'm not religious, but I can't help noticing that the religious vision of immortality is much the same as mine. It has the same two faces.

Politics and Society

8 July 1970

Girl friends and girlfriends

One of the worst disadvantages of our society is the almost complete isolation between the sexes. I can have Nat and David and who knows else as close friends, but not Wendy. And if I have a girl friend, she has to be a girlfriend. All through childhood, at least, people are allowed only to associate, in a way that is at all close, with one sex. We are cut off from half the nice people in the world.

(Age 14)

2 February 1972

Class and clothes

Hypothesis: lower class becomes middle class at the point where work clothes become more formal than leisure clothes.

(Age 16)

6 April 1984

Voting in a population of millions

In this election year I have been reading the familiar quadrennial laments that if only more American blacks would see how important it is to vote, what a difference it would make!

This opinion is fascinating because in actuality, the poor black who does not vote is behaving more rationally as an individual that the wealthy white who does. The effect of one person's vote in a national election is so negligible that voting really serves no purpose except a psychological one. The actual situation at the individual level is that wealthy whites are more tuned in psychologically to voting, not that they have a truer understanding of its importance.

On the other hand obviously the group of wealthy whites as a whole benefits politically from its higher proportion of voters. Blind luck? Of course not. As a group we have found ways to persuade each other to be psychologically tuned in to voting. My mother's horror when she heard I was not registered for the latest New York primary is the sort of microscopic act out of millions of which a social pattern is built. Her behavior can be best explained by the principles of natural selection: groups which find ways to make mothers act like that end up on top.

29 November 1988

Bribes and legal fees

By American standards, many countries are appallingly corrupt. To get your parcel cleared by a troublesome customs official in Panama, you slip him $10; he expects it. Bigger services entail payments in proportion.

This kind of thing is shocking to Americans, yet we have a strangely similar phenomenon in the institution of legal fees. For a routine divorce, you pay a lawyer $500. Closing a condominium purchase requires $1000, and to get your Green Card from the Immigration and Naturalization Service, unless you're marrying a native or have time to burn, you'd better set aside $5000.

In principle, you don't need a bribe in Panama or an attorney's fee in America to get what you want from the government. But both systems are sticky, and move very slowly or not at all if you fail to nudge them properly. All over the world, power concentrates in the hands of professionals, and they manage to make themselves expensive.

16 July 1995

The Jetsons

The children are upstairs watching *The Jetsons*, an old cartoon series I remember well from the 1960s about a family living in a gee-whiz future of robots and computers and rockets and jets.

Put yourself back in the sixties. How would you have imagined that this TV show would look thirty years later, in 1995? The obvious answer is: technologically quaint. No program built around envisioning the technology of the future, one would have imagined, could fail to look dated after thirty years.

It's curious to note that in fact, *The Jetsons* does look dated, but not for that reason. The robots and jets look as acceptable now, in a cartoon sense, as they did then. But the sex roles! The way George Jetson treats his wife, and the jokes he makes about her spending habits at the shopping mall! It's not sixties technology but sixties sociology that makes this show embarrassing.

25 January 2003

Honors and prizes

With honors and prizes, positions and such,
We must give them to those who don't want them too much.

26 May 2004

Compulsory voting

In Australia voting is compulsory. Australians accept this as part of the landscape. They don't point to it as something special and Australian, and if you ask them what other democracies have the same regulation, they don't know. Some are in favour of compulsory voting on the grounds that everybody should vote, and some are opposed on the grounds that it's better if only people who care vote.

I am in favour, on theoretical grounds. As I discussed in one of these notes many years ago, for an individual in a large population, voting is essentially irrational, for one's influence is infinitesimal. Yet society benefits greatly if people vote. It follows that incentives should be put in place to make it rational!

Non-Australian democracies depend on people to act irrationally. Well, to be precise, voting may be rational for many individuals, but not for the officially sanctioned reason that it helps them secure their interests. Rather, it's a social matter: my friends will think less of me if I don't vote (and I may think less of myself, too). But a system shouldn't be designed to depend on such clandestine forces.

The main reason I care is not that this is bad design. Rather, it's because of the ugly self-delusion involved. In the current situation, the western world pretends that voting is rational, when it isn't. Every little debauch of truth weakens us a little.

16 January 2007

Mobile phone instruction manual

I've just acquired a new mobile phone and I'm reading the user's guide. The first nine pages are devoted to safety information, alerting me to the following risks:

Risks of human exposure to radio frequency (RF) energy
Risks of usage near hospitals and other health care facilities
Risks of usage on aircraft
Possible interference with pacemakers, hearing aids, and other medical devices
Danger if used while driving
Danger if placed near an airbag (which might fling the phone at you)
Risk of explosion at gas or petrol stations
Risk in other potentially explosive atmospheres (grain elevators, etc.)
Risk of accidental triggering of explosions in blasting areas
Dangers of use if your phone or battery has been damaged
Risks if you put it in a microwave oven
Other risks associated with batteries and chargers: short circuits, heating etc.
Choking hazards
Glass parts
Risk of triggering seizures and blackouts
Risk of repetitive motion injuries

26 May 2009

Tall trees as an explanation of capitalism

Something has troubled me about capitalism. If the market is in equilibrium, then how can anybody make a profit? Isn't the whole idea that market forces drive profits down to zero?

This puzzle came to mind while I was hiking last weekend with Marc Servetnick in the Olympic Peninsula rain forest. These giant evergreens go up in 100 feet of leafless trunk before finally spreading out into greenery. It's a beautiful sight, and a fascinating phenomenon to think about. Why in the world does a tree rise to these towering heights? The whole point, of course, is to make it taller than its neighbors so it can catch sunlight.

And so we get a model of capitalism. Yes, it's a competition, with pressure toward equilibrium. Yet if he keeps moving, one player can be ahead of the pack and profit accordingly. The advantage is transient, but may last a long time.

People may quarrel whether competition for money produces a high level of goods and services, but I think we can agree that competition for sunlight produces tall trees.

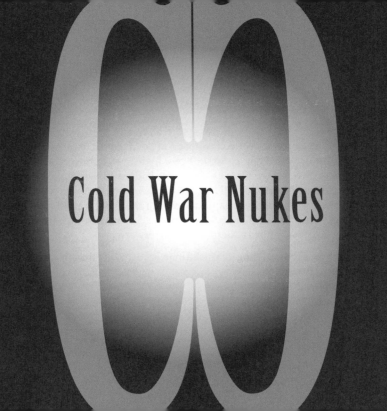

8 April 1982

St. Louis and Sverdlovsk

Here's a proposal for a short story or novel to read as our memories of Hiroshima and Nagasaki fade. A terrorist group composed of *good guys* is determined to avert a global nuclear holocaust. They construct two hydrogen bombs, and manage to hide them in two U.S. and Soviet cities — say, St. Louis and Sverdlovsk. They alert the governments that the bombs will be exploded but do not represent attack by the other side. Then the bombs are exploded simultaneously. A million people are killed in each city. The U.S. and the U.S.S.R., horrified, break previous deadlocks and for the first time begin to make reductions in nuclear arms.

I believe that such an event might give the human race a substantially better chance of survival to, say, AD 2100.

25 April 1983

Expected time to holocaust

I believe that as things are going, each new year brings between a 1% and a 10% chance of the nuclear holocaust. If this is true, our civilization has an expected survival time of 10–100 years. A rough guess might be 2½% change of holocaust per year, hence expected survival time 40 years — two generations.

I consider it useless to try to reach such estimates by carefully estimating all the relevant factors. The facts we have to go on are so hazy that such an endeavour inevitably becomes a smokescreen for the author's intuition. Therefore I offer the number 2½% with no attempt at proof: it is an unsupported assertion, but deeply felt.

This estimate is absolutely fundamental to my obsession with the nuclear peril. If I believed the risk were only 0.1% per year, the issue would subside for me into the great clutter of merely urgent matters.

Do you disagree with my estimate? If not, yet if like many people you disagree with my conviction that this issue eclipses all others in urgency, what do you think is the mistake in my argument?

25 April 1983

What's wrong with extinction?

If I am vaporized tomorrow in a painless flash, what harm has been done? If the species becomes extinct this afternoon, so what?

Jonathan Schell confronts this philosophical question head-on in Chapter 2 of his marvelous book *The Fate of the Earth*. He struggles with it valiantly. At one point he even reinvents a variant of Pascal's wager: since extinction would mean the termination of infinite generations of future humans, then so long as there is a nonzero probability that our efforts may avert the catastrophe, we are mathematically obliged to put aside all other merely finite matters.

I believe that Schell's attempt is hopeless, that no valid argument can prove extinction to be an evil. But it's a sorry human being who finds this conclusion very comforting.

1 May 1983

Seatbelts, handguns, cigarettes, nukes

The nuclear danger is one of several which, in my opinion, most people underestimate because the threat is too distant and probabilistic. Three others are smoking cigarettes, not wearing safety belts in automobiles, and the prevalence of handguns. The three mistakes shorten one's expected lifetime — very roughly — by something like five years, one year, and several months, respectively.

At least with cigarettes and safety belts, an individual can remove the threat from his own life, and not suffer for the short-sightedness of others.

Our misjudgment of the nuclear issue is graver in three ways. First, we do not have the daily spectacle of people dying from nuclear weapons, so our awareness of this peril is weaker than of any other, relative to its true magnitude. Second, that magnitude is greater: nuclear weapons are shortening my expected life by a matter of decades, not years or months. Third, the holocaust raises the specter not just of cutting our own lives in half, but of eliminating the lives of future generations.

27 November 1983

Leaders are followers, too

Nathaniel disputes my belief that reducing our stock of nuclear weapons would lessen the risk of a holocaust. Whatever the number of missiles, he argues, the holocaust will come if and only if our leaders decide to fire them. So what good are arms reductions?

I believe this argument is fallacious, for it denies the psychological connection between the size of the arsenal we maintain and the likelihood of our deciding to use it.

At present, an enormous nuclear establishment flourishes and grows, employing a hundred thousand people whose daily business it is to prepare for nuclear war. At the center sit our leaders, habitually counting silos and megatons in the tally of military assets, worrying when these numbers seem lower than the enemy's. Sometimes they claim that despite all this, a nuclear war is unthinkable. Yet inevitably they must believe this with only a part of their minds, for the existence of such active preparations for war makes anyone — even me — suspect there might be some legitimate purpose in them. How can we be sure which part will prevail in the heat of a crisis?

20 May 1984

Political exhaustion

During the past year I have joined a number of organizations opposed to nuclear weapons — I send in $10 or $20 when an appeal comes in the mail. As a result I am on many mailing lists. Two months ago I started putting anti-nuke mail aside in a special pile. Here are the organizations that have contacted me since then:

- Union of Concerned Scientists (2)
- J. of European Nuclear Disarmament
- Institute for Policy Studies
- The Golden Gate Alliance (2)
- Gandhi's Voice
- New Society Publishers
- Greenwich Village Coalition Against Nuclear Arms (3)
- SANE (2)
- Freeze Voter '84 (2)
- The Amicus Journal
- The Nuclear Times (3)
- Ground Zero
- The Federalist Caucus

When I contemplate this mass of paper I feel helpless and discouraged. Surely this fragmentation of what might have been a cohesive anti-nuclear movement has weakened it. It has unquestionably weakened my own involvement. Though I have exceedingly strong opinions on this issue, none of these organizations has captured my imagination or loyalty, for there are just too many. This evening I threw the pile away.

23 September 1985

AIDS and nuclear war

The U.S. is in the midst of a horrifying AIDS epidemic. This disease appeared only around 1980, but the number of fatalities has *doubled each year*. This year it is 5000.

In reaction, AIDS has suddenly, this summer, become a top news story. We hear about it daily. The panic has arrived, and will get steadily louder so long as the exponential growth remains unchecked.

I cannot help comparing this threat to that of nuclear war. In a hypothetical sense, the two are similar in magnitude, for if the doubling continues, AIDS will eliminate much of the human race in twenty years, just as nuclear weapons may do. But what a difference in public attention! Already one hears as much about AIDS in the news as about nuclear arms; think of the disparity three years from now, if the epidemic continues to spread. One could scarcely imagine a function of time better adapted than this to excite sustained public response: annual doubling of the death rate.

And so it is that humanity will most likely not be decimated by AIDS. Many more will die, but frantic research will probably lead to a solution in a few years. Horrible as it is, the AIDS crisis is a normal one. Nuclear war remains the abnormal, invisible crisis that we never panic about.

3 August 1987

Nuclear weapons and nuclear power

I believe that each of us is incomparably more likely to be killed by a nuclear weapon than a nuclear power plant, because if a nuclear weapon explodes anywhere in the world, it may very likely be part of a global holocaust. This difference in probabilities seems so obvious that it would hardly be worth mentioning — except that so many millions of people disagree.

For those who don't like probabilities, here is another, indirect argument that by rights should sow doubt in their minds. Nuclear power plants are designed to produce electricity; nuclear weapons are designed to kill people. This difference in design is so elementary that we easily fail to remember it! One has to believe it is a strange world indeed, if despite the best efforts of generations of scientists and engineers to make our nuclear arsenal deadly and our nuclear power plants safe, they still add up to comparable levels of risk.

5 January 1989

SDI and contraception

For a Strategic Defense Initiative anti-missile defense to serve much purpose, it would have to be almost perfectly effective. At a cost of hundreds of billions of dollars, what's the use if it stops only 95% or even 99% of the incoming missiles?

Yet we appear to have a familiar example of nearly perfect effectiveness — contraception! In each episode of attempted fertilization, the attackers number in the hundreds of millions, and yet most of the time pregnancy is prevented. Doesn't that amount to an awesome success rate of something like 99.999999%?

Not in the least, for two reasons. One is that in reality, most of the sperm don't get near the egg anyway — it's really hundreds of thousands to worry about, not 10^8. The other is that the physical situations are utterly different. Neutralizing sperm with rubber barriers and poisonous chemicals is effective and, as it happens, practical. It would be quite another matter if we had to shoot them down one by one.

26 August 1976

Continuous assessment

When the questions of where to send me to high school and university were under discussion, my mother used to say proudly "Of course, a truly good student will flourish no matter where he is put." I thought I had left this nonsense behind, but now I've rediscovered it at Cambridge University. Undergraduates here, whose success is measured solely by two sets of examinations during their three years, decry to a man the steady pressure of American "continuous assessment". There's plenty to recommend this point of view, but *not* the argument they go on to cite in the next breath: that Cambridge men are so well motivated that they will work equally hard with or without the stimulus of regular testing. I've seen both systems at work, and this claim is simply false.

3 January 1980

The literary bias in education

It seems to me that literary activity is emphasized in the school years to an excessive extent. In my English classes at Shady Hill the assignments were usually to write poems or stories, not little essays discussing my opinions or anything else. The central goal was artistic. History classes narrowed the spectrum of assignments to just narrative; they did not compensate by encouraging much analysis. When I started writing *ideas* on index cards, that seemed an activity rather unlike anything the school might encourage or be interested in.

Admittedly, children have a natural interest in stories which it is convenient to work with, but they have opinions too. As adults, they will typically spend a good deal more time with *Newsweek* than with any creative literature. In teaching them language skills, which no one would deny is the main purpose of making them read and write, why do we leave the impression that use of language is essentially artistic rather than analytical?

27 June 1981

In praise of easy learning

Some people think that stiff challenges are the best device to induce learning, but I am not one of them. The natural way to learn something is by spending vast amounts of easy, enjoyable time at it. This goes whether you want to speak German, sight-read at the piano, type, or do mathematics. Give me the German storybook for fifth graders that I feel like reading in bed, not Goethe and a dictionary. The latter will bring rapid progress at first, then exhaustion and failure of resolve.

The main thing to be said for stiff challenges is that inevitably we will encounter them, so we had better learn to face them boldly. Putting them in the curriculum can help teach us to do so. But for teaching the skill or subject matter itself, they are overrated.

7 November 1993

Good test takers

A familiar complaint against the reliance on tests in our educational system is that the tests measure not just knowledge of the subject but also an artificial skill called "test-taking ability". It isn't fair that one student may score twice as high as another when they know the material equally well!

I agree that this phenomenon exists. I am not sure, however, that it is all bad. What is a good test-taker? In large part, surely, it is someone who remembers to look at the big picture, who habitually tries to figure out what the problem is aiming at rather than just focusing on the details. Now in the "real world" that one reaches after school, the same skill of seeing the big picture is decisive. Two attorneys may know the law equally well; two professors may know their research fields equally well. If one sees the big picture and the other does not, who will end up with the notable achievements?

Thus a case can be made that the test-taker phenomenon, far from being a device by which an unforgiving system contrives to reward artificial skills over genuine talent, is just the opposite.

3 July 2000

How Oxford gets it backwards

I've read somewhere that in the early days of electronics they got it exactly backwards. Communication was done over wires, while entertainment was broadcast on radio waves. It has taken a century to reverse the arrangement so that we converse on mobile phones and televisions benefit from the high bandwidth of cables.

Oxford University has two functions in a similarly inverted configuration. The intellectual creativity of its scholars is a university's most precious resource, and at any American university, it is utilized via a basic principle: the man or woman teaching the course has a fair degree of autonomy, ability to innovate. Meanwhile it hardly bears mention that most professors don't waste their time admitting freshmen.

Here in Erewhon, oops I mean Oxford, the situation is exactly reversed. Because of the centralized examination system, scholars are required to teach to a committee-mandated curriculum and every impediment is put in the way of innovation in the classroom. Meanwhile fellows give a week of their lives each December to admitting next year's students, a huge effort with dubious results. How you can teach effectively, they insist, if you haven't hand-picked your students?

8 August 2001

Bits of information in an Oxford degree

A student leaves university in America with a transcript full of information. Even with grade inflation, there are thirty marks of A or A− or B+ or B to look at, each one attached to a different course like Advanced Calculus or 20th Century Philosophy or Introduction to Economics. Grade-point averages are constructed from these transcripts and reported to three digits of accuracy.

An Oxford graduate finishes with no transcript, just a degree result which may be a First, a II.1, a II.2, a Third, a Pass, or a Fail. Failures are more or less nonexistent, and the numbers last year for the other degrees were 691, 1925, 374, 39, and 3, respectively. The corresponding probabilities are 23%, 63%, 12%, 1%, and 0.1%.

If you add up these probabilities times their base 2 logarithms, all times minus one, you find out how much information there is in an Oxford degree. The result is: 1.37 bits of information.

20 June 2005

Quality vs. quantity in British academia

British universities are ruled by the Research Assessment Exercise, a system that increases the competitive pressure greatly and is probably, I suppose, for all its drawbacks, a good thing on balance. But one feature exasperates me about the RAE. In each assessment period — the latest is seven years long — we are allowed to put forward only four publications on which to be judged! The official line is that what matters is quality, not quantity. This is nonsense. No leading university would appoint a professor who published so little. In fact, both quality and quantity matter — and among those whose importance was enhanced by having plenty of both one might mention Newton, Einstein, Shakespeare, Dickens, Bach, Mozart, and Picasso.

An analogous myth is put forward at the high school level. Across England, universities admit students based on three A-level scores. You can take 4 or 5 or 6 A-levels, and perhaps that will impress someone, but the official line is that that we care about quality, not quantity — potential, not achievement. What nonsense to deny the importance of achievements below age 18! What a waste to press young people to excel in just three subjects! The effect on Oxbridge admissions is terrible inefficiency and inequity, for there are innumerable applicants who get three A's, so we can't use A-Levels to distinguish them. I ache to change Oxford's procedures to increase that number from 3 to 4.

17 January 2007

Is private schooling immoral?

There's a fuss this week because another government minister has revealed that she sends her son to a private school. In England, no issue is more emotive. Many feel deeply that this is wicked. A Balliol fellow has written a book on the subject.

I believe that in general, there are many cases where it is appropriate for a government, or a society, to pressure people to do one thing where they might privately choose another. For example, I am in favor of regulations discouraging the use of gas-guzzling or highly polluting cars.

When it comes to schooling of children, I feel strongly the other way. One of the strongest of all human drives, indeed in a sense *the* strongest, is to get the best for your children. This is the engine that has powered much of the progress of civilizations through history. This English class-obsessed objection to private schooling takes the view that parents' desire to get the most for their kids is selfish and wicked. Well, I don't think a society that takes this view is going to thrive for long.

13 August 2007

Clinton and Blair at Balliol

Both Clinton and Blair listed Balliol as their first choice Oxford college, and both were turned down. So Clinton went to Univ and Blair to St. John's.

What passions are linked to this little story! So I found in discussing it recently with Kate. To an American, it's an amusing reminder that admissions committees make mistakes. Obviously, in American eyes, if Balliol had judged better it would not have rejected these two titans. But many Oxford eyes, or perhaps English ones, truly see it differently. Let us agree to ignore matters of publicity and fund-raising and consider the question of principle: from the point of view of its central mission, should a top college or university strive to admit the leaders of the future? Americans cannot imagine that one could seriously doubt it. But the "English" view is that politics is orthogonal to the real business of a university, which is intellectual, so no, we should not care in the slightest whether or not we grab the Clintons and Blairs. I know that many of my Balliol colleagues feel this way.

I find this view bizarre in the extreme. If our goal is to foster the use of reason in the world, then surely getting our hands on future Presidents and Prime Ministers will help foster it. Or is our mission solely to foster the use of reason among academics?

Britain

6 January 1998

Electrocution in British bathrooms

British bathrooms are cunningly designed, by law, to enhance the probability of electrocution.

I became aware of this anomaly the other day when the light bulb blew in our bathroom here in Oxford. In any other room of the house, the light is on when the switch is down and off when it is up. Then when you are changing a bulb, it's a trivial matter to turn off the juice.

British bathrooms, however, are not allowed to have the usual switches, on the theory that it is too awful to contemplate them being touched with wet hands. Instead they have cords that pull from the ceiling. Each pull toggles the state of the light from off to on or back again.

Of course, by the time you get around to changing the bulb, you have pulled the cord a number of times in your frustration and have no idea what state the fixture is in. Maybe 0 volts, maybe 240. Good luck!

26 February 2000

A conspiracy of comma splices

Email at the office: *"Shirley won't be in today, she has a tummy bug."*

Sign in the car park: *"Do not lean cycles against this wall, use cycle racks, cycles will be removed."*

Poster at the railway station: *"Great Western staff are here to help you, our staff are entitled to work in a safe environment."*

From a Balliol governance paper: *"We received seven applications from MIT, one subsequently withdrew."*

In America you rarely see this kind of punctuation; it must be zealously stamped out by schoolteachers. Here in England the custom is different. Comma splices are as common as bicycles, though they are usually avoided in books and newspapers.

I have asked a dozen people about this difference between the countries, including Balliol fellows as senior as Jasper Griffin, Oxford's Public Orator, a savorer of words whose wife is American. But I have not found one who agrees that there is a difference. "That punctuation is wrong," they say; "educated people don't do that." I yearn to share my observations; they dismiss me as an ignorant colonial.

16 September 2001

British pronunciation of American names

On British radio and television, one comes to expect certain mispronunciations:

MARY-land instead of *MERaland*,
PENtagun instead of *PENTaGON*,
PoTO-MACK instead of *PoTOmick*,
Los ANgeLEEZ instead of *Los ANgeliss*.

I've encountered some more exotic specimens too, such as

...an event in the Californian town of "Pay-lo Alto"...
...the conference took place "on Rhode Island"...

These errors amuse me, but what's bizarre is how much they also annoy me. Damn it, I feel when I hear another Mary-land, why are these British so stupid! It's a foolish reaction, and I try to fight it. Of course, I also try to understand it. It seems the drive to divide the world into "us" and "them" is deep, and we react with emotion to the little markers that help us discriminate.

22 May 2005

Baseball and cricket

In a baseball game a team gets 27 outs, around one every three minutes when they're at bat, and scores maybe five runs, roughly one every twenty minutes. It's those rare events, the runs, that bring the fans to their feet and determine which team wins. Dash and flash are what make the difference, and that's why runners are always straining for second base and fielders throw the ball like a bullet at their teammates across the diamond.

In test match cricket the figures are reversed. In a day of play a team scores hundreds of runs, one a minute, each greeted by polite applause. It's the outs that bring the roars and cheers, for there are at most ten of them, often less than one an hour. What marks the truly great batsman is not the sixes he wallops but his ability to hit away for hours and never get out. Defensive batting is what makes the difference, and that's why runners take few chances and fielders make few fast plays. Gloves might come in handy, but they're hardly indispensable in this game that is fundamentally negative.

28 November 2006

Towels in Japan, kickstands in England

Here in Japan, it seems incomprehensible that bathrooms have sinks but no towels. Why are there no towels?, I asked a Japanese friend. Well, we don't really need them, he said, since most of us carry a handkerchief, and he pulled out his own handkerchief to show me. Nevertheless I note in men's rooms that in practice some Japanese men wash their hands, then shake them a little and leave the room damp. Ludicrous!

It's easy to see foreign cultures as sillier than one's own, so I asked myself, what is a comparable foolishness in the West? An example that came to mind is kickstands on bikes. Every bike in Japan has a kickstand, but in England or the U.S., half of them don't. So every time you stop your bike, you have to find something to prop it up against. I've never understood why people put up with this nonsense, but they do.

20 November 2006

American and British: the advanced quiz

I cannot resist recording some of the differences between US and UK English I've enjoyed over the years. Only somebody who's lived on both sides will know all these.

In the US (easy): sidewalk, trunk, hood, fender, cell phone, sneakers, panties, pants, in a week, in the hospital, make a decision, value, take-out, study, smart, fresh, arugula.

UK equivalents: pavement, boot, bonnet, wing, mobile phone, trainers, knickers, trousers, in a week's time, in hospital, take a decision, value for money, take-away, revise, clever, cheeky, rocket.

In the US (harder): macaroni and cheese, sorted out, can I help you?, part (in hair), bangs (likewise), envision, get a word in edgewise, drunk driving, training wheels, cup of coffee, sled, whine, oriented, to tell the truth, thumbtack.

UK equivalents: macaroni cheese, sorted, can I help?, parting, fringe, envisage, get a word in edgeways, drink driving, stabilizers, coffee, sledge, whinge, orientated, to be honest, drawing pin.

And then there are the pronunciation differences, like vineyard, condom, dandruff, Renaissance, scenario, Lithuania, weekend, macramé, hurricane, geyser, glacier,....

23 February 2008

Famous Oxford

I think Oxford (pop. 130,000) may be the most famous city of its size on earth.

What are the competitors? An obvious possibility is Cambridge. However, my Chinese friends tell me Oxford has wider name recognition among the Chinese billion.

Another possibility is Bethlehem. Every Christian has heard of it. The Chinese know Oxford better, but I should make inquiries among South Americans and Africans.

26 April 2009

More time Catholic or Protestant?

Has England spent more time Catholic or Protestant? Of course it's an artificial and ill-defined question, but let's do our best. Suppose we say that the country was created in 1066 and was Catholic until 1533. In addition there were five later Catholic years under Queen Mary, so for the sake of the tally we can imagine that England was Catholic until 1538. That's 472 years. Add 472 to 1538, and you get 2010. Yikes! Is next year the one in which the balance will shift to Protestant? Can somebody do this calculation more carefully and tell us the particular date on which to mark the transition?

28 April 2009

Kublai Khan and the age of Oxford

When I bring people to lunch in Balliol, I like to tell them that the college was founded a generation before Kublai Khan led the Mongols against Japan. (The Divine Wind or Kamikaze beat them back.)

Here's another window on the age of Oxford. The faculty of my Numerical Analysis Group are fellows of seven colleges: Balliol (1263), Exeter (1314), Oriel (1326), New College (1379), St. John's (1555), Pembroke (1624), and Worcester (1714).

30 October 1980

Who was John Stuart Mill kidding?

John Stuart Mill claimed that he never put aside even the smallest question until he had solved it completely to his satisfaction (*Autobiography*). He recommends this regimen as conducive to intellectual rigor. I have always been incredulous at these remarks. Did interesting questions occur to Mill no faster than he could resolve them? If so, his mental processes are alien to me. Was he kidding himself, or us? If so, it's a cruel joke. The principle is noble, but living up to it is beyond human capacity.

24 January 1981

B. F. Skinner

B. F. Skinner is an example of someone completely focussed on one idea to the exclusion of all others. This comes out in reading his recent *Notebooks*. The single-mindedness of his efforts there to explain all things behavioristically has to be read to be believed.

Such one-sided thinkers pose a fascinating problem. For the world as a whole they are a natural and a good thing — obviously behaviorism will get a better hearing if 10% of us are entirely committed to it than if each of us is 10% committed. The whole of Skinner is one part of the balance that makes up our world.

But can you imagine *being* B. F. Skinner? What balance can there be in his world?

20 February 1990

Antonio Salieri and old furniture

It is sometimes lamented by those admiring, say, a sturdy piece of antique furniture, that "they don't build 'em like that anymore". I expect there's some truth in this, but there's also a pinch of absurdity. Obviously anything still here after a few hundred years must have been sturdy enough to last that long! Nobody would deny this elementary selection principle, but it's easily forgotten.

Actually the selection process is a bit complicated, for the survival of a cultural artefact depends on its *value* as well as its sturdiness. In this respect paleontology is simpler than antiquing, for fossils show the hard parts of the organism only, and that is all there is to it. At the other extreme, a field like musicology that deals with cultural intangibles is also relatively simple: Mozart's music survives longer than Salieri's solely because we value it more, not because it's published on acid-free paper. Anthropology, and antique furniture, lie somewhere in-between.

Like the present, the past was full of the flimsy, ephemeral, and second-rate. For every Mozart there were a hundred Salieris.

25 April 1994

Noam Chomsky

Chomsky is extraordinarily intelligent, well-informed, and rational. Like Bertrand Russell, he's on a level above almost everybody. A Bach in a world of Handels.

Yet in the end I admire Russell and find Chomsky a pain in the neck. Why does he have to be so *negative* all the time? Russell is often negative, true, but not always; and with Russell, one is swept away by the feeling that this is a man brimming with life. Chomsky by contrast seems brimming with bile.

Intellectually, can he be faulted? Yes, with care. Many of Chomsky's seemingly outrageous claims may be literally true. But they are consistently 10% out of alignment with the best statement of the truth — skewed towards the view that all events are controlled by a conspiracy of corporations in a society ingeniously contrived to support that conspiracy.

In Chomsky's world, anybody with any power is evil. Well, OK. By a certain definition of evil, that may be true. But then, I say, you have not 10,000,000 instances of individual evil to excoriate, but a general phenomenon to explain and counteract. Stop berating the 10,000,000 and propose something positive!

22 October 1995

Greg Sheehan, Bill Gates, and Napoleon

Freshman year at Harvard, 1973–74, I lived with two roommates in Wigglesworth B-11. One of them, Greg Sheehan, had a French mother and was a fan of Napoleon. We had a world map over the mantelpiece, and late of an evening, Gregoire would put on his best French accent and tell us how he planned to conquer the world just as Napoleon had done. We will attack here!, then here!, then here!, he would say. I will make the world again un grand empire!

Twenty feet to the east that year, another freshman was living in Wigglesworth A-11. His name was Bill Gates. Curiously, in the time since then Gates has largely fulfilled all of Greg Sheehan's promises. Before age forty he has become the richest man in the U.S. and one of the most powerful in the world. Like Napoleon, he is brilliant and driven, possessed of infinite energy and multiple talents — military and administrative in that case, technical and commercial in this.

Gates is at an age at which Napoleon was at the height of his powers, though a few clouds had gathered over Spain and Portugal. Five years later, the empire was defeated and Napoleon was in exile on Elba. We shall see if Microsoft manages to do better.

17 April 2000

Shoeless Joe Jackson

One day eighty or ninety years ago, a baseball player called Joe Jackson played barefoot because his shoes caused trouble with the blisters on his feet. For the rest of his life, and in memory long after, this man had a nickname he couldn't shake.

Keep Shoeless Joe in mind as you imagine your own life's impact on the world. You may be noticed for the smallest thing — and for the smallest thing, you may be noticed.

15 April 2002

Piet Hein's grooks

For reasons I cannot distinguish,
Piet Hein could write Grooks in English.

For reasons not hard to explainish,
I cannot write Grooks in Danish.

25 May 2004

Jesus Christ's curriculum vitae

We all like to do well, to be respected by friends and strangers and remembered when we are gone. And we all know that there's a lot of randomness in how these goods are distributed. A bit of luck can make all the difference. I could tell you about a guy who's no better than I am, but ten times as famous....

Perhaps it is instructive to consider the case of Jesus Christ. This man lived about 2000 years ago and was probably very impressive. I don't know how ambitious he was, but undoubtedly he had special gifts and deserved to do well. But to be turned into a god worshipped by billions of people for thousands of years? It's a safe bet that this level of recognition is all out of proportion to Christ's actual talent and, well, if I were one of his competitors, it would just eat me up.

1 August 2004

The hedgehog, the fox, and Beethoven

Isaiah Berlin, if I remember right, had a theory along the following lines. Some people are hedgehogs, who know one true thing, and others are foxes, who dart around after many smaller things. Tolstoy's problem, according to Berlin, was that he was a fox who wanted to be a hedgehog.

We were listening to the Eroica in the car yesterday, and as always, I enjoy Beethoven but find him frustrating. The tunes are marvellous — he was so clever with melodies! But Beethoven, unlike Bach, is never content to give into a tune and go with it. Instead his ideas shimmer in a perpetual dance of feints and fadings and profound observations from the timpani. He would, of course, be contemptuous to hear himself described as "clever with tunes." Beethoven fancied himself much deeper, and that's what's wrong with his music. Like Tolstoy, he wanted to be a hedgehog.

22 October 2007

Thomas Edison the physician

Thomas Edison visited the Eiffel Tower the year it was completed, 1889. There's a brass plaque at the top commemorating his visit, which calls him

> *Thomas Edison, the American physician and inventor.*

That's a nice double error. They mean physicist. And Edison was no physicist.

28 September 2008

Strang, Knuth, and Lax

Three of the most impressive people I know are Gil Strang, Don Knuth, and Peter Lax. Each of these world-famous mathematicians has a perception that is simply at a higher level than other people's. And all three give outstanding talks.

Yet these heroes of mine have a curious point of style in common: all of them, when giving that talk, act a bit bumbling and helpless. Poor Gil can't quite get his head around the mathematical point he's trying to make, it seems — is there anyone around who might have some ideas? Poor Don can't quite finish a sentence, such a struggle — can anyone help? Poor Peter is such a kindly gentleman, so courtly in that old European way, but he can't quite begin to put together his thought at all — is there anyone in the audience please who could lend a hand?

The trickery can be annoying, but boy is it effective. Strang and Knuth and Lax get just where they were aiming by the end of the hour, and you're on the edge of your seat. Is the bumbling unconscious? Intentional? A symptom of genius? The frailty of older men? Should I, too, learn to hesitate and swerve when I talk?

Optimizing Your Life

13 June 1970

Marriage vow

When I get married, I want to be careful. I am going to make absolute, absolutely sure that whoever I marry is someone I can live with. I don't want anyone I will end up hating, or have a lot of arguments with. I want someone I really love and can be happy with for a hundred years. Whenever I say to myself I've found the girl and I want to marry her, I hope I remember to wait another year or two and then think about it again.

(Age 14)

16 October 1970

The paradox of effort

Does it require "work" for me to do my homework? Was there any "effort" involved in, for example, getting good grades with Miss Caudill? No. My mind was so set on working, and is now on doing homework, that an alternative is not conceivable. I know today that by next Tuesday I will have written a theme for English; therefore, why worry? There's no effort required when I know it will be done. Just wait around for Tuesday night, and, lo and behold, the theme will be done, just like magic.

(Age 15)

25 November 1983

A watched to-do list never boils

I perpetually make lists of projects that have to get done — paper to write, old friend to look up, problem to solve, wedding gift to buy. Yet they never get done, it seems. Day after day I consult the list and see there are items to add but none to cross out.

Then a month passes, and I run across the old and forgotten list. Suddenly I discover many of those tasks have been finished off while I wasn't looking! Apparently what keeps me from spending time on what I ought to be accomplishing now is always the greater urgency of what I ought to have accomplished last month.

26 May 1986

Zenogrook

With an infinite series of errands to do,
How can we ever get bigger things done?

Here's the secret: the series is infinite, true,
But it sums to a number just smaller than one.

2 March 1987

Probabilistic analysis of my commute

Walking to the subway stop fascinates me from a probabilistic point of view. The trains arrive unpredictably at random intervals. So long as you are ignorant of when the next one is coming, there's no need to hurry — the expected effect of arriving one minute earlier is merely that you'll catch a train one minute earlier. But everything changes as soon as information becomes available. If you hear a rumble, or see the train crossing the bridge over the Charles — time to run! Now one minute can be expected to save you five.

I enjoy walking at a relaxed pace. In fact, I enjoy it so much that as I round the final corner from which I can see that bridge over the Charles, I am tempted to avert my eyes lest I see a train approaching and have to hurry. But of course, that's not playing fair with the probabilities; the calculation wasn't based on self-imposed ignorance. If you avert your eyes, the cost in average commuting time will be a real one.

4 September 1987

Not touch typing

Millions of capable people spend their days at computer terminals without knowing how to touch type. What a waste, to hunt and peck for a lifetime at 30 words per minute! With a computerized typing instructor, ten hours' effort would suffice to make anyone touch type at that speed, and now that manual keyboards are as obsolete as horses, it's an effortless advance from that point up to 60 wpm a few months later. But people won't invest the ten hours, so they end up throwing away ten thousand.

It is a marvel that society tolerates this waste so complacently. The victims are often unconscious of their inefficiency, or proud of it. How would these same people feel on the highway if they were obliged to drive at 30 mph all life long while other cars streaked by at 60?

26 February 1999

Why life is exhausting

For many of us, life seems overwhelming and exhausting. So many things to attend to! We must feed the children, take out the trash, replace the light bulb in the hallway, prepare tomorrow's lecture, call a plumber about the radiator, put gas in the car, clean up the kitchen, get a new front brake on the bicycle, sew a button on the red striped shirt, buy new socks, on and on and on.

I have figured out where the trouble lies. We are trying to perform parallel tasks with a serial processor. The secure, easy way to attend to a thousand things is *simultaneously*. One subsystem notices if the car needs gas and solves that problem; another notices if a shirt needs mending and mends it; and so on. Indeed, this is just how our bodies work: there are thousands of matters of daily maintenance to worry about, but each cell contains the full instruction manual and they're all switched on to attend to their duties simultaneously. You need not worry that your body will say, darn it, I forgot to fix the scrape on your knee, I was so busy with that bruise on your elbow! Maintenance is easy in parallel, absolutely exhausting if you try to simulate the same effect by running from one task to another.

14 December 1999

Oubliette

I have a spot where papers go
If maybe I'll need them but probably no.
I call this box my oubliette,
And I haven't needed one yet.

3 January 2000

Advice for the young

A couple of years ago I gave a series of lectures for numerical analysis graduate students. I finished with six items of advice:

1. Subscribe to *Nature* or *Science*.
2. Pay attention to people.
3. Work on problems that interest others.
4. Work on problems you enjoy.
5. Become skilled at MATLAB.
6. Learn to touch type.

10 January 2002

Bondage is easier than discipline

Sometimes it helps to constrain ourselves artificially. If you're worried about your weight, it's best not to have goodies in the refrigerator. For me, endless email is a temptation, so I've arranged to have access to it only at the office. At home, I am forced to do other things.

What's going on here? If we are rational creatures, why are many choices worse for us than fewer? Why is freedom slavery?

My favourite example is the little chain that ties a pen to a table. There's no mystery about why a shop or a library might want to tie its pens down, but a pen chain is useful even at home by the telephone — even if you live alone!

7 February 2004

Workaholic

I'm a workaholic. I listen to the news while shaving, read *The Economist* while playing piano, and get up early to shuffle papers even when there is no deadline pressing. I couldn't bear to eat breakfast alone without a book.

Curiously, the quality of that book and its value to my further development are not so important. In fact one can't help admitting that on a moment-to-moment basis, the essence of being a workaholic is not that one must be getting worthwhile things done, but merely that one must be *constantly busy*.

Now you might say, this shows how foolish workaholics are. Why, they even drive themselves when there's no purpose! However, there's more to it than that. A workaholic doesn't ask at each moment, shall I or shan't I? He or she simply works. It is the *habit* that is all. To consider shall I or shan't I at each moment might sound optimal in theory, but for actual human beings, it would be exhausting. In the end, much less would be accomplished.

14 July 2005

The advantages of an optimal life

Like many busy people, I've developed all kinds of strategies to optimize my efficiency. Such a tricky balance of how much time to spend on email, paperwork, errands, social life, housework, research!

Now as we know, any smooth function looks like a parabola near its bottom. This means that if you miss the optimum by a small amount ε, your loss in efficiency is much smaller, of order ε^2. What a blessing! It means that day-to-day, you need not worry about doing a little too much of this or a little too little of that, for the penalty will be just quadratically small. And so optimality brings peace as well as efficiency.

22 January 2006

Physics, psychology, and Petter Bjørstad

As graduate students at Stanford we all travelled by bike. I didn't think much of Petter Bjørstad's affectation: he kept his ten-speed in the top gear up hill and down, never changing gears at all. Of course Petter knew the equations of force and distance and work as well as I did, and I thought he was rather a show-off to ignore them.

Twenty-five years have passed, and Bjørstad and I are successful professors of numerical analysis. Outside the office, I am an ordinary middle-aged man with an ordinary physique. Bjørstad meanwhile has climbed 459 mountains, including Mount Whitney, Mont Blanc, Kilimanjaro, Aconcagua, and 85 others over 4000 meters.

The Life of the Professor

19 March 1981

Academic distractions

The natural thing to focus on in doing research is the *problem*. I am suspicious when I talk to somebody who seems focussed on anything else. Some of them are obsessed by the literature, and can't talk without beginning each sentence with the name of a colleague. Some are obsessed by the formalism, and dazzle you with technical issues that aren't the real issues. Technical precision is necessary, and may turn out to be the most difficult part of the research, but rarely is it the heart of the matter.

2 May 1981

Zooming in, zooming out

It is vital to see the big picture, to know where your work is going. Some of your time must be spent as a philosopher. And it is equally vital that you cheerfully roll up sleeves and work hard at minute details, for hours or days at a time, a business nearly impossible except when the philosopher in you is willing to sleep. Just as our eyes can handle a range of many orders of magnitude in brightness, our minds must perform at every level of detachment to be maximally productive.

12 March 1987

My office at MIT

I have a small but efficient office here at M.I.T., and since I spend a great deal of time in it, I have developed hundreds of little tricks to use the space most efficiently. Special places have been settled upon for typewriter, workstation, colored pens, magnetic tapes, paper towels, pads of paper, letters to mail, coffee machine, stapler and tape dispenser, squash racket, seminar announcements, computer manuals, hand lotion, radio, reprints, old journals, textbooks, lecture notes, cognac, extension cords,....

The usable volume is about 10' by 15' by 8' — 1200 cubic feet. In the course of the year, I spent about 2–3 hours in the office for each of those cubic feet.

29 December 1994

Academics vs. politics

Like many other academics, I find that with the years, I shoulder more tasks of an organizational nature in addition to research. Other people are generally involved, and so I find myself engaged in politics and fascinated with its complexities.

I am coming to believe that there is a fundamental conflict between intellectual and political life. Without too great exaggeration it can be put as follows. The essence of the intellectual life is the impulse to analyze everything indiscriminately, to understand the explanations for everything. The essence of political success is the contrary habit. Whatever one's internal understanding of the world, one must train oneself never to express those conclusions needlessly, for they may offend. Truth must be offered up by teaspoons, always with a purpose.

To be a successful politician, it seems to me one has to proceed either by a kind of dumb instinct, without any too explicit internal model of the world, or by duplicity, raising a wall between one's thoughts and one's utterances. I do not mean to suggest that the latter is entirely shocking; after all, it might be considered just an extreme development of the art of politeness. My guess is that the average parliament or congress is inhabited by both sorts, but here I am speaking beyond personal experience.

5 September 1997

Shredder

I've just spent my first five minutes in my new office at Oxford. In the corner I found a document shredder.

I'd like to try this out, I thought. So I put in a page from a pile of papers for recycling, and it shredded it nicely.

I felt a bit foolish about this little experiment. How childish to shred a piece of paper with no sensitive information on it, just for the fun of it!

Fortunately, nobody would be able to tell.

28 August 2004

The diameter of intellectual space

Scientists like to add new things to our body of knowledge. And so it is that over and over again, authors present their contributions without pointing out links with others'. Sometimes they know the links and don't mention them. More often they don't know them, or know them fuzzily, and consider making connections a low priority. Indeed, clarity itself is a low priority, for is it not a sign of lack of depth?

In extreme cases these irresponsible papers make our scientific edifice worse, not better, all the while adding to their authors' reputations for depth and importance. In reaction, when I write an article, I aim for it to "shrink the diameter of intellectual space." It should make connections, make the reader's world a little more coherent.

This is a moral issue, and it has cost me, for I am sure my obsession with clarity has encouraged my colleagues to regard me as a shade less deep. In 1990 at MIT, if I had been less clear as a teacher and writer, I believe I would not have been voted down for tenure. Does this sound farfetched? Consider that the Dean, in explaining MIT's decision, informed me that promotion required a world-class research record — clearly assuming since I was a good teacher that I must think otherwise!

29 November 2006

Ten themes of how I do research

My modus operandi as a researcher has some strong features. Here are ten.

Typing. Doing things fast at the keyboard is crucial; how do non-typists manage?
Writing. I am obsessive about writing, and do it in tandem with the research.
Getting to the bottom of things. I can't rest till I understand a problem to the bottom.
Computing. I hardly ever do anything without computer experiments to guide me.
Seizing on a mystery. Something unexpected — that's where progress comes from!
Internally driven. I do what I think is important, not much guided by fashion.
Methods driven. I'm driven by the methods more than by the applications.
All over the place. I work in half a dozen areas. I am hard to classify.
Graphics. I can't understand things until I can draw pictures and make plots.
History. I consider it immoral to discuss a topic without connecting to predecessors.

Of course I am proud of a combination that has made me successful. Yet in looking at this list, I note that these admirable traits are not universals. Plenty of good people don't care about typing, writing, computing, or graphics; plenty are driven by applications; plenty have a clear single focus. But I think every scientist should be obsessed with getting to the bottom of things, and should seize upon a mystery.

23 July 1970

One of my greatest wishes

One of my greatest wishes in life is to be with people who are musical. I never have been.

(Age 14)

22 September 1970

Invention or discovery?

To invent is to create from one's imagination; to discover is to make known for the first time.

Is a musical tune an invention or a discovery?

I think most people would say an invention. I would lean towards discovery, however. I tend to think of the number of basically different tunes as finite, and that to write a really good one is more to discover it than to create it.

(Age 15)

12 March 1980

Music vs. speech

Here is one proof that music is not symbolic in the same way that speech is: to listen to a news program we just turn on the radio; but we play music soft or very loud, according to mood.

17 June 1981

Randomness in art

In all kinds of creative efforts a little randomness can be invaluable. When I improvise at the piano I make careless mistakes, and also play random notes intentionally — then I hear them, enjoy them, and build them into themes, and my music is better for it. The explanation is that our powers of interpretation are enormous, while our facility to create out of a vacuum is relatively weak. So any kind of input, even random, may be enough to stimulate us. The same should hold in oil painting, poetry, cooking, you name it.

The artist may build on the random input himself, or he may leave some randomness in the product. This is then something for the consumer's imagination to work with.

Error is not the only source of randomness. Another is the conflict of constraints that cannot all be satisfied. In poetry and song lyrics, the need for rhyme or meter forces the writer to introduce a random element in his verses, form the point of view of the meaning, though he may not like to admit this. If he has the wrong instincts, the result seems foolish. If he has the right ones, it may be sublime — so that the constraints of form, far from asphyxiating meaning, may nourish it. Bob Dylan is a genius in this respect: his verses often border on gibberish, yet come across as profound.

23 April 1995

1D, 2D, 3D musical instruments

Musical instruments fall in two great classes: 1D and 2D/3D. Strings are of course one-dimensional — violins, pianos, guitars, harps. So, essentially, are woodwinds and brasses — flutes, organ pipes, trumpets, bassoons. For all its undulations and varying bore diameter, the essence of a saxophone is a 1D resonator. The same goes for the human instrument, the vocal tract.

Percussion instruments, by contrast, are usually multidimensional. Drums involve a 2D membrane, sometimes coupled with a 3D kettle. Bells and cymbals are 2D. So are castanets, more or less. And so on.

The significance of the two classes is that 1D resonators tend to have their eigenfrequencies related by simple ratios. Thus they have the potential to sound harmonic to a degree that 2D and 3D instruments generally cannot achieve. Drums sound like drums, simplistically speaking, because the zeros of a Bessel function are not related by simple fractions, unlike the soldierly progression of eigenvalues for a stretched string.

A curious intermediate case are the 1D percussion instruments. Think of the triangle, or the xylophone. Sure enough, these sound much more "musical" than the others.

16 August 1995

Pure and applied music

Pure vs. applied mathematics — it's like classical music vs. rock and roll. Many of the best people go the classical/pure route. The standards of performance are higher; indeed, the levels of talent are awesome. It commands more respect from the people who matter. Its practitioners have far more knowledge of the foundations of their subject.

Yet something goes wrong. All that talent is not properly channelled. Classical music and pure mathematics get trapped in their own history, their own self-awareness, their own high standards — and what do you find? Outstanding musicians and mathematicians who don't do much exciting and have little impact. In the end it is the rockers and rollers in both fields, looser and sloppier but with their heads properly attached, who do much of what's really important. Meanwhile their superiors grumble and know this isn't fair.

19 August 1995

The future of the cinema

Technology, a case can be made, killed the art of painting. Yes, artists still do beautiful things with watercolors and oils. But since the arrival of photography and then color photography, it isn't the same. Art isn't central to culture anymore.

Technology has something to do too with the current parlous state of music. In the last two centuries, orchestral music was central to culture, and it produced lasting works of genius. In the first two thirds of this century, jazz and blues and rock and roll produced their own works of genius. But recently, something has gone wrong. Electronic devices now produce every sound imaginable at the touch of a keyboard — and the music is worse. On the radio, thirty years later, we still hear the Beatles and Bob Dylan, because nobody nowadays is so creative. I think this is in part because technology made it all too easy. There is no style in tennis without a net.

Now what about movies? In my opinion they have been great all century long — they were great in the 1930s and they are great now. But is mediocrity destined to arrive in the next few years? Since 1994 or thereabouts, digital editing of video images has become routine. Reality can be bent and blended at the touch of a keyboard, and the effects so far are astonishing. The net is down. What will happen to movies?

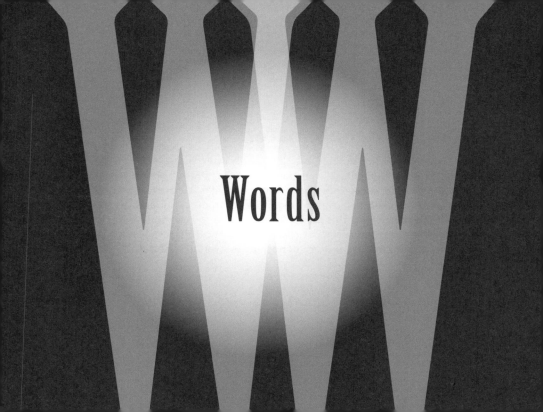
Words

8 September 1976

Common denominator

Most clichés have a respectable past: if you only listen with a fresh ear, they express an imaginative thought or a clever analogy. But the phrase "common denominator" is an embarrassing exception. I hear statements like "Whatever their differences, Charlotte Bronte and Jane Austen share the common denominator of interest in love and marriage." How this kind of usage relates to the common denominators of arithmetic is beyond me. Most clichés were conceived by a thought now sadly dormant, but this one seems born of nothing but verbal confusion.

14 August 1987

Unspellable and unpronounceable

Let us say that a language is *pronounceable* if one can determine the pronunciation of a word from its spelling, as a rule, and *spellable* if one can determine the spelling from the pronunciation. All four possibilities are readily found in nature:

	pronounceable	unpronounceable
spellable	SPANISH	HEBREW, ARABIC
unspellable	FRENCH, GERMAN	ENGLISH

Much as I cherish the richness of history that is preserved in the idiosyncratic spelling of English, not to mention my own rich investment in learning all those idiosyncrasies, I believe we would be better off with a reform aimed at hoisting English towards the upper-left corner.

22 February 1988

Stock market meltdown

We recently lived through a stock market crash, but it seems half the headlines described it instead as a "meltdown." That word didn't even exist in 1929, at the time of the last crash!

There's a process here akin to osmosis. To describe events we reach for analogies from other fields, and words from more familiar areas tend to diffuse into less familiar ones. These days nuclear disasters figure larger in the public mind than financial disasters.

11 February 1989

Which animals are also verbs?

You can crow at a friend, ferret out an enemy, bug a room or an acquaintance, ape a politician, outfox him, or otherwise cow him. You can dog your opponent, horse or monkey around, worm your way in, chicken out, or rat on a schoolmate, and for that matter, man the oars or people a room (with a little help from your friends).

Of course bear and fly and bat are verbs too, and so is flounder, but those are examples I won't swallow.

26 August 1997

An irregular singular

According to my dictionary, the singular of *timpani* is *kettledrum*.

30 August 2000

Stet

Walter Gander, from Zurich, is writing a book, and I was editing a chapter. After marking one of my corrections, I changed my mind. I was about to write "*stet;*" but would Walter understand that? To play it safe I wrote "original is OK."

Walter later confirmed that no, he didn't know the word *stet*. (My English-German dictionary translates it as *Es bleibe stehen*.)

I guess Latin is only used these days in Britain and America. Luckily, you can communicate with educated people all over the world in English.

9 September 2003

Trefethen record adjectives number

Each morning here at the University of Queensland, I have occasion to use four adjectives in a row. I go to the coffee cart and ask for a "tall skinny flat white."

No worries.

7 August 2006

Wizards and Wallahs in Windows

I plugged something into the computer and Windows popped up a screen saying "Welcome to the Found New Hardware Wizard."

There are lots of other Wizards on my computer too: the Accessibility Wizard, the Network Setup Wizard, the System Restore Wizard, and the Files and Settings Transfer Wizard, for example, not to mention the Web Publishing Wizard and the Cleaner Management Wizard and the Scheduled Task Wizard and the Forgotten Password Wizard.

It's a great usage, but Microsoft needn't have invented it, for doesn't the Indian word "wallah" do the same job? We should really speak of the Forgotten Password Wallah. Maybe the Indians already do.

13 May 2007

Caution: race horses

We passed a truck on the motorway yesterday with a sign on the back —

        CAUTION
      RACE HORSES

Funny how different a construction that would have been if it had said —

        CAUTION
         LIONS

22 December 2007

Centripetal like I

Sloppy English speakers say "me" instead of "I", so among the more refined, a counter-error has grown up of saying "I" where it should be "me". Between you and I, Baltimore is boring. Would you like to have dinner with Diana and I?

An upwardly-mobile friend of mine produced the platinum version of this mistake the other day. We were talking about things whirling around and being driven outwards. "It's a matter of centripetal force," he said.

10 April 2009

Paste in the taste

Last night's pizzeria menu was full of the hilarious errors of English that you see in unpretentious restaurants in Europe. Grouped under headings including "Entrants" and "Dry paste of hard wheat," we found dishes like these:

"Salad of chicken with pineapple & pink sauce"
"Raviolis of meat in the taste"
"Spaghetti sea & mountain (robbed with mushroom, ham, shrimps, & aromatic species)"
"Provolone of cheese in hot small pan, with leaf of basil (with small toasts)"

We particularly enjoyed the stern words at the end of the menu, which I report in full:

"*The VAT is included in the prices*. Is only issued for box and group, IS NOT CHARGED SEPARATELY. You ask for the note the waiter and pay to himself and don't get up to pay the box, THANK YOU. WE WAIT FOR YOU IN THE NEXT VISIT TO BARCELONA."

You might think such chaos places us on the edge of verbal breakdown. The truth, wonderfully, is just the opposite. These people live in a mélange of Castilian, Catalan, French, Italian and English and can hack their way successfully through any encounter. The silly menu is actually a badge of their linguistic ease.

Writing and Literature

10 June 1979

Good and bad writing

I could go on forever on the associative nature of thought — I have belabored it in previous cards and will not stop with this one. One more indication of the way we think is the great difference between good and bad writers. Part of it is linguistic skill, yes, but no writer is great unless he also has the knack of stirring up associations in his readers' minds with every phrase. Shakespeare is sublime for just this reason, not because of the keenness of his philosophy. Really, it's an embarrassment to the human psychology that Shakespeare is as good as he is.

21 September 1980

On being edited

Almost any change in the wording of a sentence causes a change in its meaning. I think this is one reason why certain people resist the emendations of an editor so stubbornly. If you expect changes in sound or syntax only, your instinct is to fight any change in sense as a proof of the obtuseness of your editor. The trick is to recognize that small changes in meaning are to be expected, but are not necessarily destructive. Most of these semantic fine points weren't part of your idea to begin with; they only appeared when you set the idea in words, and can usually be replaced by other fine points without materially affecting it.

29 January 1983

Literary allusions instead of analysis

Reading Dyson's *Disturbing the Universe* makes me realize how suspicious I am of literary references and allusions. I feel their power. The right connection to a work of literature adds depth to a matter, making one sense the complexity of its connections to other things.

The trouble is that too easily, recognition of the literary context becomes an excuse for lack of analysis. Having widened the discourse, one senses one has accomplished something. Indeed one has, but it's a dead end. That kind of thinking does not built a structure that further ideas can be grounded on.

Harping on the logical skeleton of an argument seems unsophisticated, and as the years go by I do less of it in these index cards. But ultimately we are too sloppy thinkers to deserve the luxury of leaving logic implicit. Our culture goes too far in its broad respect for the literary approach to thought.

19 March 1983

Novels make life worth living

Every time I read a good novel (currently *Anna Karenina*), it astonishes me to see to what extent getting wrapped up in fictional characters seems to lend a purpose to life. There are some pleasures that can come from many sources, but in my life the pleasure of feeling part of a group — which is what makes life seem to have a meaning — comes from exactly two: close friends, and good books. In the absence of the former, the latter is an alarmingly good substitute.

1 July 1983

Mathematical abstraction vs. literary art

No intellectual endeavour is more precise than the statement and proof of an abstract mathematical theorem. Allusive and high-minded literary prose lies near the other extreme, for it is highly evocative but not precise at all.

Yet in part the mathematician's abstract precision and the writer's evocative ambiguity serve the same purpose. Both are means to render an idea more flexible and therefore potentially more powerful. The mathematician abstracts in the hope that his theorem will thereby become applicable to new situations that he cannot foresee. The writer is suggestive rather than explicit so that his ideas, too, may be amplified by unforeseen resonances in the mind of the reader.

20 August 1995

Authorship as fixed point iteration

Before computers, articles and books went through one or two or three drafts before publication. Authors had to be skilled at envisioning how copy would look in print that was splattered with corrections and reorderings and insertions.

Nowadays, if the author is finicky, articles and books undergo endless revision. At each step an excellent typeset manuscript is polished so that it becomes still more excellent. It is not too great an exaggeration to say that the process works like this: the author keeps reading through the text endlessly until, at some iteration, not a single correction is made. Then the loop is exited and the text is published.

If f denotes the function that maps a draft text to its author's edited version of same, we have what mathematicians call a fixed point iteration $x \to f(x) \to f(f(x)) \to \ldots$, continuing until a point x^* is found with the property $f(x^*) = x^*$. It would be fine if we could say that this property guaranteed that x^* was optimal. Unfortunately, all we really know is that its flaws, small or large, are orthogonal to the perception of the author.

7 March 2001

Men who scribble and men who mutter

I can't organise my thoughts except by writing them down. Not for me the sequence of finishing the research, then beginning the paper. No, the thinking and the writing go hand in hand, with the paper progressing through twenty electronic versions as it helps the thoughts develop. Even for a routine weekly meeting with a student, I have often typed out half a page of notes in preparation.

I may be a little extreme in this regard, but I think nobody would regard my reliance on writing as embarrassing or deranged.

Lately, now, I notice I'm talking out loud to myself more than I used to. Strolling through Hyde Park as I am today, organising my thoughts, how natural and helpful it is to speak the key words out loud! Instant shape and focus!

Society takes, er, a different view of this behavior. I try to shut up as others pass me on the path.

29 August 2004

A mathematical model of good writing

One might think that the quality of a book or essay or poem, or at least its intellectual as opposed to artistic quality if one may make such a distinction, should be equal to the magnitude of the ideas it presents. However, there must be more to it than this, for we all know that a work does not feel great if its style is too explicit. Ambiguity, or at least open-endedness, are essential. The reader wants to be a participant, not just a spectator.

We can take this observation as the basis of a mathematical model: *the quality of a written work is equal to the quantity of thinking it stimulates in the reader.* According to this view, a pound of wisdom neatly laid out is worth less than an ounce of fuzzy figuring that the author induces the reader to do for himself. Readers want to be exercised, not informed. It's not the amplitude of the signal that matters, but how much it resonates.

Applied to conversation rather than literature, the same view seems familiar, even trite. We all know that there is nothing worse than a conversationalist who cares more about what's going on in his head than in yours.

As the years have gone by I've tried to be less clear in these notes, but it's hard work.

23 December 2007

One more for Strunk & White

I am a child of Strunk & White, and I don't pretend to have many tricks of the trade of writing that are my own.

Perhaps one is a rule that I've followed for some years now: avoid the expression "of course". Yes, this phrase may add nicely to the flow of a sentence, for those to whom the point is indeed obvious. But you can almost always imagine a reader to whom the point is not obvious, and for that person, "of course" is patronizing.

31 December 2007

The advantages of obscurity in writing

I just wrote an index card about Kate's birdbox from Uncle John. In that card I didn't draw conclusions explicitly — I didn't even mention the thoughts in my mind about God and the argument from design and the impossibility of man understanding God's motivations. It seems more powerful to leave things unstated.

Why do we do this? Why is the best writing often inexplicit, open-ended, even obscure? Here are some of the ideas Kate and I have considered.

Horace theory: even if there's a clear and simple meaning, the reader will absorb it more fully if he/she is forced to work it out for him/herself.

Shakespeare theory: there are multiple meanings, and the impact is greatest if different readers are free to resonate with different parts of the thought.

Bob Dylan or bullshit theory: there's no real meaning, just the sparklings of a kaleidoscope, and the author is clever enough to quit while he's ahead.

T.S. Eliot theory: there is a real meaning after all, but obscurity excludes some readers, adding to the value for those in the club.

11 July 2008

The appearance of inconsistency

There's a principle of ethics and public life: you must avoid not only impropriety, but also the appearance of impropriety.

I've noticed an analogous principle of publishing: you must avoid not only inconsistency, but also the appearance of inconsistency. For example, in an index, you might be tempted to list one item as on pages 32 and 33 and another on pp. 52–53. You may have a genuine distinction in mind; but still, you risk disturbing the reader.

Or in a journal article, you might want to cite four references [12,7,8,11]. No doubt you're making a point of chronology or relative importance by putting those numbers out of order; but again, be careful.

29 October 1971

I can't distinguish a book from a movie

When thinking back on some story I encountered a few years ago, I often find it hard to remember whether I read the book or saw the movie. I see before me an image of scenes from the story, but I often cannot tell whether I constructed those scenes from a verbal description, or whether I am remembering the real picture of what I saw.

This suggests to me that the memories we can recall of past events, at least of this sort of past event, are very much processed and filtered through our brains. I cannot be remembering very vividly, if I cannot even recall my source of information.

(Age 16)

9 May 1972

Looking glass memories

We fill in our memories of the past to a large extent with present-day knowledge of "what it must have been like." Thus I can remember playing baseball with David Paul several years ago, but the details of his appearance that I seem to visualize are almost certainly derived from my present-day knowledge rather than from a memory of his appearance at the time.

This was brought home to me by the memory I have carried for several years of the bus stop in Australia that my Seaforth school bus used to stop at to pick me up in the morning. I remembered this stop as on the right side of the road!

(Age 16)

21 July 1974

My brain is not nearly full

A couple of weekends ago WRKO's latest gimmick was to play in order the top hundred hits not only of the last year, but of the last *five* years. The thought of what sort of a forgotten tune might have won ninety-seventh spot five years ago, say, made me wonder at first about how good a gimmick this could hope to be. But I found to my surprise that, even five years ago and even towards the bottom of the hundred, I knew essentially all the songs. Apparently I have within me a thousand or more songs in some form of replica. That's impressive evidence of the storage capacity the brain can throw around.

(Age 18)

20 May 1984

Lifelong memories of baby teeth

The "E" key on this typewriter is loose, and my middle finger feels it wobbling around on its stalk as I type. Each time this happens I am reminded very powerfully of a similar sensation I have had before, when I was a child: feeling with my tongue a loose tooth that is soon to fall out.

The vividness of this memory is amazing when you consider it goes back twenty years. Apparently my mind will stand ready for the rest of my life to interpret the wobble of a baby tooth instantaneously as a familiar sign — an oddly irrelevant skill for a head that experienced its final baby tooth decades ago.

25 July 2007

Fox hunt

At lunch today it was driving me crazy. I couldn't remember the name of an old acquaintance whom I wanted to mention in the conversation. I kept coming up with GEOFF HUNT, but that wasn't it. I thought the first name was probably right, but I knew the second wasn't.

Darn it. Minutes later found me still puzzling.

Finally it came to me. The man is GEOFFREY FOX.

How odd, *hunt* and *fox*. Coincidence? I think so, but I am not sure.

12 September 2007

A spoonerism in Spooner's house

Since my office on Keble Road is in the building Spooner and his family lived in for 25 years, I feel an affection for spoonerisms. Today a nice one came along. Hilary Ockendon and I were talking of a new staff member and the name I came up with was

> Laura Howe.

But no, this wasn't right. After a moment we remembered that the correct name is

> Helen Lowe.

6 July 2008

A multiplicity of mnemonics

Gil Strang, though an Honorary Fellow, didn't know the combination for the door of the Senior Common Room at Balliol. Here's a mnemonic, I offered. Remember that book by Ray Bradbury, "Fahrenheit..."?

I planned on Gil saying "451", whereupon I would say yes! — and now you just have to add a 2 and you have the combination, 4512.

But Gil surprised me. He didn't know the book title. My mnemonic was dead in the water.

So I told him — 451.

Ah, well that's easy to remember, he said. That's 41 times 11!

2 February 1972

Am I awake or dreaming?

It's funny how, when dreaming, we think we are awake, but when awake there is a much stronger conviction of the fact, no doubt whatsoever. Somehow it is like skimming through a book to see where you left off reading: at several places you think you are there, but as soon as you reach the real position you are absolutely sure.

(Age 16)

23 March 1973

Alcohol and marijuana, hawks and dogs

In my few experiences with alcohol and marijuana I have found very unpleasant the feeling of being out of control. The dreaminess of my perceptions and the diffuseness of my concentration under those influences seem to contrast greatly with the apparent sharpness and alertness of normal consciousness.

It is frightening to realize, then, that my normal condition is not alertness on some absolute scale, but merely the level of alertness I am used to. My most incisive moments may be as bad as total inebriation in comparison to the comprehension of beings somewhere else in the universe, or even in comparison to some humans here on earth. Or perhaps I am near the top among humans, and what seems like alertness to other people would be fuzziness to me.

The same problem on a smaller scale is exhibited by acuity of vision. I tend to think of my normal focus as total acuity and any blurring of that as transition from full to partial perception in some absolute sense; but of course both levels of acuity are relative, and each would seem blurry to a hawk and superlative to a dog.

(Age 17)

9 July 1977

My clock has a foreign accent

Somebody poked his head into my office the other day and told me the hour hand on my clock wasn't right. It sits halfway between two hours, and he'd misread the time at first glance. Somehow I'd never noticed the problem.

Hour hands have to communicate one of twelve distinct possibilities at a given time, and in this they are like the organs of speech. The tongue in a given language is allowed only four points of articulation, say, and the listener does his best to perceive a sound he hears as just one of those four possibilities, even though the true sound may lie anywhere along a continuum. So my office clock has a heavy accent, which confused one observer into getting the message wrong.

26 September 1978

Rubber gloves illusion

I just bought a pair of rubber gloves, and I've washed the dishes two or three times with them. I still can't believe that my hands are not getting wet each time I immerse them — all the usual "wetness" sensations are fully present. You couldn't ask for a better demonstration that we sense the clues, not the facts.

9 August 1982

Y and Z on a German typewriter

On German typewriters the positions of "y" and "z" are interchanged. Since "y" is a good deal more frequently encountered, after a couple of hours I've learned to press the "z" key for "y", but not yet the other way around. In this charming intermediate state of skill I instinctively press the same key for both letters! It's another illustration that logical consistency between items in the brain is not inherent in the mechanism, but learned.

13 October 1987

Putting caps on pens

I'm careful about putting caps back on pens. My father is careless, so his pens dry out.

Maybe I'm too careful. Yesterday I went into the secretary's office when no one was there and noticed a box of a dozen felt-tip pens on the disk. The cover of the box had an illustration of one of the pens uncapped and ready to write. This bothered me. The sight of an uncapped, unattended pen was not a pretty one.

9 February 1991

Coffee pots and prejudice

Restaurants in the U.S. brew coffee into a glass pot which sits on a hot plate. Two things then happen: (1) the level in the pot goes down as cups are poured; (2) the coffee deteriorates as it burns. Thus there is a correlation: a fuller pot means fresher coffee, on average.

I find I have absorbed this correlation deeply in my perception. Full pots *look* fresh to me and emptier pots *look* stale. Though well aware intellectually that there are exceptions to the rule, I find I cannot suppress the distaste I feel at the sight of a nearly empty coffee pot.

Is this not the essence of prejudice? Some prejudices do not correlate with the truth, of course, but those are straw men, hardly worth a comment. The difficult fact is that many prejudices *do* correlate with the truth. Perception of coffee pots makes a good case study, for it is an example we can analyze without offending anybody.

18 January 1993

Car accidents and the tails of distributions

Lately it has been recognized that processes once thought gradual actually proceed, in many cases, by uneven jumps. Biological evolution seems substantially directed by rare cataclysms; gorges are carved more by the once-a-century flood than by the day-to-day trickle. The effects we care about are often controlled by the tail of the distribution.

Another example may be car accidents. My habitual view has been that the road is full of drivers like me, sober and sensible people subject to fluke errors that cause accidents so random as to be unavoidable. But lately it has occurred to me that although 98% of drivers may fit that description, most accidents are probably caused by the other 2%. Some of those 2% are the well-publicized drunks, but I'll bet there are others hidden behind their innocent windshields whose performance is nothing close to normal — some young and foolish, some old and impaired, some just lusty and irresponsible.

Imagine if we could change things so that the ordinary 98% of vehicles were painted the usual blues and silvers but the other 2% were painted bright red. Soon we would stop noticing the 98% almost entirely and give all our attention to the red devils. Working from a better model of the road, we would be better drivers, no longer making the error of confusing the distribution with its mean.

25 July 2007

Is this a man or a woman?

I am listening to a talk by a Chinese mathematician, W. Is this person male or female? I can't tell! The name, the voice, the dress and the shape, to my ears and eyes, are androgynous. No jewelry, indeterminate haircut. Exasperating!

Ah, now I see it. Men's shirts button one way, women's the other. From this fact, after inspecting my own shirt, I calculate that W. is a woman. And suddenly my perception shifts and it seems obvious, and I wonder how I could have wondered.

Knowledge and Truth

21 June 1976

Where astonishing things come from

There are two ways that astonishing things may come about: either someone invents them, or they pop up naturally in some cauldron of statistically varied muck and their astonishingness floats them to the surface. Murders in television dramas are of the first sort, murders headlined in newspapers the second. The idea that God designed the living things on earth is of the first sort; the theory of evolution by natural selection is of the second.

5 September 1978

In defense of jargon

Classifying things is not understanding them, of course, and when I listen to Prof. Bryson naming wildflowers in pictures without hesitation, dozen after dozen, I can't help thinking: what sort of knowledge are these names? Yet always I conclude YES, this is worthwhile knowledge, and I envy Bryson his expertise. Moreover, I know from experience that when I approach a discipline for the first time, my appetite for the terminology is hearty. However much specialized jargon may make a field obscure to the outsider, I do not believe it can be dispensed with, nor even wish it could.

Names are not ideas, but there is no better way to fix a structure of ideas than by pinning it throughout with names. Truth as we humans can deal with it is inescapably bound up with language, and the best we can do given this limitation is to learn plenty of definitions.

31 January 1979

On the value of philosophy

Philosophy *is* worth thinking about, even talking about. Philosophical questions matter, and each of us has a richer life for trying to come to terms with them.

But philosophy is not worth arguing about too seriously, at least to the extent that one's intent is to make real progress towards the truth. Most of the philosophy believed in 2100 AD will be based on scientific developments which have not yet unfolded — so state your views by all means, but don't haggle over them. The ostensible purpose of professional philosophers and their institutions, namely to advance the field, is pretty much a vain one.

10 June 1979

The better part of validity

I have often marvelled at the fact that I began writing these index cards simply as a list of thought patterns that kept recurring in my head, to exorcise them. Very soon the subject matter became ideas, and the goal to preserve them.

Yet this transition was not so great as it seems, for the distinction between a well-worn train of thought and an idea is tenuous. Our thought processes are fundamentally associative, not deductive. A good idea is often not much more than a sentence from one's internal monologue that sounds particularly good, and familiarity is the better part of validity.

19 July 1980

Discussion, opinion, and truth

To John Stuart Mill and to Socrates, for example, the intellectual life is centered about *discussion* and the weighing of conflicting *opinions*. The generation of a new idea is a rare event compared to the eternal business of comparing and judging all the old ones. Progress in understanding, they believe, is won through scrupulous standards of reasoning adhered to during endless talk.

I hope that as a greater proportion of our knowledge becomes scientific, we are growing out of this narrow view. Though controversy exists in science, it is by no means its core. Though discussion is important, other things are more so.

Some loss is dealt the life of ideas by such a change, for the devaluation of vigorous discussion is a movement away from subtlety. The loss must surely be more than balanced, however, by a corresponding gain in the solidity of our knowledge. Mill seemed uncertain whether any truly permanent increase in knowledge is ever achieved. A hundred years of scientific progress later, one cannot doubt that it is.

25 December 1980

The itch to enumerate

Although ostensibly one states a generalization in order to make the enumeration of its particular applications redundant, nevertheless I have the urge when writing to work out all the applications of each idea. This runs me up against space limitations.

The itch to enumerate is not really unreasonable, for various reasons. But being conscious of it can help when you are pruning your writing or suffering its pruning by others. Tell yourself that the very omission of an application is a mark that your idea is fruitful and reliable!

3 October 1982

Oft thought, but e'er so well expressed?

Many of the thoughts in these index cards have undoubtedly been put in writing before, some of them perhaps many times. I have a fantasy that one day, by magic, I will find on my desk a book of *Trefethen's Duplicated Index Card Notes*: one page for each of my cards, and below it on the same page, that text from another author of any era which best matches it. How many of these pairs do you suppose would be essentially perfect duplications? What a feeling this book would give one of the continuity of human thought through the ages!

19 January 1986

Final causes and instability

Obviously final causes operate around us. Species advance with time, market economies encourage efficient methods of production, and so on. But to explain a final cause we must always reduce it to efficient cause mechanisms.

Often the mechanism involves a phenomenon of instability. In a world of primitive organisms, the slightly more advanced one will leave more descendents. In a market of inefficient producers, the slightly more efficient one will profit disproportionately and grow.

In physics a simple final cause is often minimization of energy. In principle you might tip over a jar half-full of water and find that the water stayed at the top, held up by air pressure. But of course that's not what will happen in practice, for if the interface between air and water deviates ever so slightly from a perfectly flat surface —

4 January 2003

How can so many things be so important?

In music, there are a thousand tunes that seem so beautiful and natural, they are surely great discoveries, not mere inventions. In painting, there are a thousand graphic ideas so pure, so elegant, that when we enjoy them they eclipse all others. In literature, there are a thousand turns of phrase so apt that when we contemplate one of these gems, it seems to capture not just a little of the world but a great piece of it. In science, there are a thousand discoveries so important that they explain, why, just about everything. And in writing one of these index cards I routinely have the feeling, *this* is one of the truly big principles that govern our world, one we should all be continually mindful of, the index card more important than all the others.

I think the essence of this wonderful phenomenon, which makes space for all our lives to be interesting, is combinatorial. There are hundreds of ways in which you can string two words in a row, or two musical notes, or two brush strokes. Increase two to ten, and the hundreds become billions. One in a million of those billions is going to have some real meaning. So there you have it, thousands of beautiful creations which are so so so short! They are jewels!

30 January 2003

Calibrating a colleague

One day last October one of our newly arrived graduate students told me they'd had three meters of snow the night before in his home town in Latvia! I was impressed at this remarkable news from a country I've never visited.

Not long afterwards he mentioned that his mother speaks seventeen languages fluently. I no longer think it snowed three meters that night in Latvia.

An optical illusion

21 January 2005

Here are a bunch of dots. On the left, we can't resist perceiving them as vertical lines; and on the right, as horizontal.

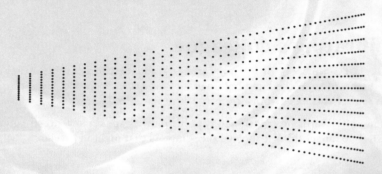

You couldn't ask for an illusion more elementary, more easily explained. Yet isn't it a prototype of so much of our world? Our notions of this and that are *right*, and we couldn't do without them. But alter a few parameters smoothly to other values, and those notions may suddenly and quite rightly jump. Remember this illusion next time you think about the nature of reality, or next time you disagree with a friend.

16 March 2009

Philosophical rigor and molecular physics

The driving questions for many philosophers are What can we know for certain? How can we reason with rigor? Yet when you read their writings about these matters, it's hard to take them seriously. Any clown can see that nothing is certain and no reasoning is infallible. A lot of hot air! We clearly know a great deal in practice, some of it pretty solidly, but it's silly to imagine we can know anything with certainty.

My own view of the matter is founded on an analogy to molecular physics. My living self is ultimately just a confederation of molecules bouncing about. Just as my temperature and density are not exactly defined properties (they are only approximations that work well because I contain so many molecules), so my life, and my thoughts, and logic itself, do not precisely exist either. At bottom there are no living organisms or thoughts or logical proofs — just bouncing molecules. Of course in some sense I am alive, and I have a temperature, and I have thoughts, and I try to reason, but none of these things really exist in an ultimate sense.

When a philosopher argues about a delicate point of logic and knowledge, the project is as doomed as asking, What *exactly* is the temperature of this glass full of water? When *exactly* does life end if you swat a fly?

23 December 2009

Will the sun rise tomorrow?

Hume formulated the problem of induction: how can we know that the future will be like the past? If the sun has been observed to rise on each of the past million days, what principle allows us to infer that it will rise tomorrow?

The answer, I believe, is the Copernican Principle of J. R. Gott. The argument is one of Bayesian probability, and it goes like this. Suppose you observe a system at a random point in its lifetime. The probability that you observed it in the last 1% of its life is 1%; the probability that you observed it in the last millionth is one in a million.

So imagine humanity has seen a million sunrises in a row and that's all we know of the matter (i.e., forget our knowledge of solar system physics). Might the sun fail to rise tomorrow? Yes, certainly; but that would require today's sunrise to have been one-in-a-million special. The probability that the sequence of sunrises will terminate tomorrow is on the order of 10^{-6}; sometime next week, $O(10^{-5})$; next year, $O(10^{-3})$.

I think this argument captures pretty well the psychology of how we actually interpret our world. Jacob and I were discussing it on a ski lift, which stopped while we were in mid-air. After 20 seconds immobile, was it reasonable to expect it might start up again at any moment? Yes, of course (and it did). What if we had been immobile for 20 minutes?

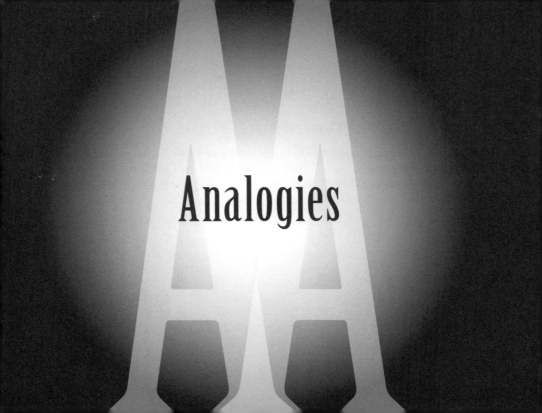

2 February 1972

Morality and swimming pools

Getting up from bed is like diving into a swimming pool; one holds back from doing it, even knowing that when it is done one will be happier. Logically we know that jumping forthwith out of bed or into the pool will make us happiest, yet we don't do so.

In Samuel Butler's terms, getting up and diving in are greatly moral acts, and I find I have much more respect for those that do.

(Age 16)

19 July 1975

Antonymy, dissonance, nonsense, humor

Two words that are antonyms are, of course, really very similar. "Hot" and "cold", for example, are both adjectives that describe temperature. Only in what they assert about temperature do they differ. This is a common property of interesting contrasts: the contrasting objects differ not fundamentally, but in a significant small way.

Other examples: if you combine together two random sounds, you hear a third random sound; but if you mix very similar sounds you hear dissonance. In language, nonsense is not random noise at all, but words in sequence that come close to without quite reaching meaningful speech. In the wider world, a scene that is almost what one might expect but not quite will have a touch of humor, if the difference is of a certain sort. Another scene that distorts a few key features of reality appears surrealistic — while something vastly different from reality would only appear abstract.

Thus the five phenomena of antonymy, dissonance, nonsense, humor and surrealism are brothers. No doubt I have overlooked some other members of the family.

(Age 19)

13 June 1976

Nature vs. nurture of societies

If you ask the nature vs. nurture question about a society rather than a single person, you get the old question of whether individual men have much effect on history.

8 July 1978

How analogies work

Analogies are implicit abstractions: they reveal essential qualities by a kind of differencing process, without being obliged to describe them explicitly. In this freedom lies their power.

2 June 1986

Meaning, communication, fitness, survival

In linguistics there is a thorny problem of *semantics*. What is the connection between the structure of an utterance and its meaning? Defining "meaning" turns out to be nearly impossible. Some linguists have retreated to the view that not much can be said about what an utterance means beyond specifying the circumstances in which it is spoken.

In biology there is an analogous problem of *fitness*. Species obviously tend to become fitted to their environments, but it is very hard to specify what this fitness consists of. One is quickly reduced to the seemingly vacuous definition that the traits that constitute fitness are those that lead to survival.

This analogy sheds light an old criticism of Darwinism. The critics say that natural selection cannot explain the development of advanced organisms, because it is a tautology: what survives is what's fit, but what's fit is what survives. Here is an equally valid argument that language cannot achieve communication: what's spoken is what has meaning, but what has meaning is what's spoken.

I believe these apparent paradoxes can be resolved as follows. True, the only *exact* indicator of fitness is survival. But this does not prevent fitness from *correlating* with other traits such as strength, speed, and intelligence.

12 January 1988

Evaporation and the brain drain

Evaporation of a liquid works like this: the molecules are bouncing around at random with various speeds, and some of the faster ones have enough energy to escape the liquid entirely. So they fly away, leaving the remaining molecules a little less energetic on average. In other words, the liquid becomes cooler.

Brain drains work the same way. The intellectually or economically livelier particles in America's inner cities, in the north of England, or in parts of Asia and Africa tend to fly away from those surroundings that have too little to hold them. As a result, a simple matter of statistics, the culture they leave behind becomes intellectually or economically cooler.

2 September 1989

Trickles, streams, and floods of money

I think of money as a system of rivers and tributaries flowing through the population. How much you get for yourself depends on how skilful you are at scooping up what flows by, yes, but also, inevitably, on the size of the waterway you live beside. An ordinary worker with an ordinary salary — say, a teacher or a truck driver — lives on a steady but modest river not so far from the source, and must scale his financial ambitions accordingly. Meanwhile even a mediocre corporate lawyer can expect the money to pour in by the tens of thousands of dollars, camped as he is by the Amazon.

12 January 1992

Why is a bus like a microwave oven?

Because they're both most useful when you're on your own. Taxis are expensive if there's just one or two of you, and conventional ovens are slow. But if you're in a party of three or four, the balance shifts and the constant beats the linear function.

14 January 1993

Good language and good wine

I've often noted that knowledge of wines seems almost to narrow one's pleasures — so many bottles fall short! Being ignorant of wines myself, I am fortunate that I can enjoy all kinds of vins de table that a connoisseur wouldn't share a room with.

Devotion to the precise usage of language is much the same. To those who are sensitized, certain exquisite pleasures are available, but so much of ordinary usage is so painful! The only difference I can see between fine language and fine wines, apart from the circumstance that fine wines are expensive, is that, ahem, I myself happen to be devoted to the one but not the other.

Rather than abandon my love of language, or this analogy, I conclude that oenological expertise, despite the downside, probably makes for a better life and is something I should set about acquiring one of these days.

16 July 2003

An analogy should be imperfect

To be useful, an analogy should be good but not perfect. If A is *exactly* like B, then they are the same and there is no point in comparing them. Comparisons become fruitful when A is 90% like B. Then the analogy serves as a magnifying glass, revealing details of that interesting 10%.

This reminds me of musical instruments. We think of a guitar string or a flute as a resonator with a set of eigenvalues, that is, natural frequencies. Strictly speaking, however, such an analysis imagines that the instrument is a closed system — and if that were so, we couldn't hear it! A useful musical instrument must be a good resonator, but not perfect.

Of course, musical instruments aren't *exactly* like analogies.

10 February 2004

Purple monkey dishwasher

You know that game of Chinese whispers, where A whispers a sentence to B, who whispers it to C, and so on until it comes back distorted to A again? Matt Burrage claims that at the end, one always gets the phrase "purple monkey dishwasher."

If so, this is standard laser physics. In a laser, an arbitrary initial signal bounces back and forth between the mirrors at the ends of the cavity, soon settling into the shape of the dominant eigenmode. Whatever the initial excitation may be, you always end up with purple monkey dishwasher.

6 January 1979

Where babies come from

Once when I was a child I felt a gap in my understanding and asked my mother, "How come only married women have babies?" I had supposed that babies grew automatically at random intervals in the stomachs of grown-up women, and couldn't see why marriage should have any effect on this process.

My mother's response was inspired. "As a matter of fact, sometimes unmarried women have babies too." With this answer my world fit together perfectly again.

This story strikes me as revealing a great deal about how difficult it is to acquire knowledge in this world — scientific or otherwise. Recognizing that there's a question to be answered is difficult to begin with; recognizing that a possible answer is not quite perfect is much more so. Possible but incorrect answers abound even when no mother is wilfully misleading you. We grown-ups are only quantitatively more acute than children in noticing what answers don't quite fit, and often we lack entirely the data which should reveal the disparity.

27 June 1983

Why don't Americans wear seatbelts?

Cigarettes may be more deadly, and nuclear wars terminal. But for undiluted idiocy, not wearing seatbelts when you drive is in a class by itself. It is preposterous to argue that the feeling of freedom you may get from not wearing them is worth the death risk you face. Yet 80% of Americans ride unbelted, and 150 of us die each day from car accidents.

Now of course there are valid explanations for this phenomenon at all sorts of levels, and one shows naiveté by declaiming that it is astonishing that two hundred million people behave so stupidly. But an explanation is not a justification. I maintain that the nearly universal decision not to wear seatbelts is one of the marvels of our time.

28 June 1986

Bad logic in a good cause I

Women's lib is founded on two facts,
Serene and stately:

That women are the same as men,
And that they differ greatly.

1 July 1986

Bad logic in a good cause II

The anti-nuke folks also grind
Two most distinguished axes:

They swear the bombs will kill us all,
And also raise our taxes.

17 November 1996

Math prerequisites for public office

In the television debates between President Clinton and Senator Dole that preceded the election just past, the question was asked, what about public disenchantment with voting? Mr. Dole, what would you like to say to citizens who think that since they have just one vote among millions, their vote doesn't make any difference?

They are so mistaken! answered Mr. Dole. Why, one vote certainly can make a difference. I've seen this happen many times in the Senate.

5 May 1999

Achieving world peace

In the middle of NATO's terribly destructive but so far unsuccessful bombing campaign in Kosovo, a sign has appeared in the window of Oxford's Friends' Meeting House at 43 St Giles':

Bombs don't solve problems.

I read that and thought sadly, yes, there's much truth in that.

Then I read the sign in the adjacent windowpane:

Prayer and creative peace-making do.

10 August 1999

No need to wear a bike helmet

Here's a proof that there's no need to wear a bike helmet.

Suppose you go out on the bike, wearing that helmet. Suppose, as is almost always the case, you don't have an accident. Then, of course, you didn't need the helmet.

On the other hand suppose you do have an accident. Well now, life is chaotic — sensitive dependence on initial conditions, butterfly effect and all. It follows that if you had not put on the helmet, you would have departed some seconds earlier, encountered different traffic, etc. — a whole different trajectory through the strange attractor of life. And thus, virtually certainly, you would not have had an accident.

18 March 2001

The last bug

Jon Chapman and I had a good-natured discussion. In his major work on threshold exponents for transition to turbulence, he admitted he'd found an error that changes the results.

I expressed concern. In such a long paper, if he's found one error, well, I'm only human. This is going to diminish my confidence that what remains is error-free.

Ah, no, he explained with a twinkle. Surely if I've corrected an error, the article is now more correct than it was, not less!

18 July 2002

A theorem of E. O. Wilson

E. O. Wilson is a great naturalist with a big heart, but I think he's forgotten his calculus. On p. 29 of *The Future of Life* (2001) he writes

> The worldwide number of children per woman fell from 4.3 in 1960 to 2.6 in 2000.... When the number of children per woman stays above 2.1 even slightly, the population still expands exponentially. This means that although the population climbs less and less steeply as the number approaches 2.1, humanity will still, in theory, eventually come to weigh as much as the Earth and, if given enough time, will exceed the mass of the visible universe.

In other words, we have

Wilson's Theorem. *Every monotonically increasing function diverges exponentially to infinity.*

16 August 2004

Do needles inject drugs?

According to doctors, tens of thousands of drug users in the US have become infected with HIV because they were unable to obtain clean needles. Our conservative government has a policy of not providing free needles, on the theory that having needles at hand encourages drug use.

Notice how perfectly this view contradicts that other conservative dictum: "Guns don't kill people, people kill people."

9 December 2009

Non-monotonic logic on British Rail

I took the 7:51 yesterday to London via Reading.

Outside Reading, the train stopped and sat still for some minutes. The announcer on the loudspeaker explained: we've arrived in Reading ahead of schedule and are waiting for our platform to be free.

Eventually we got to London, 15 minutes late. The announcer apologized and explained: this was due to the delay in Reading.

God and Religion

28 June 1970

There is no God

As far back as I can remember, I have never doubted that there is no god or anything like it. I say this only as a record, not because I feel the point is questionable. I have three excellent reasons:

1. There is no evidence that there is a god, and as time goes by things which have been attributed to gods are all explained elsewise.
2. The existence of a divine power is logically ridiculous. Who created him? How does he operate? None of the powers attributed to a god make sense.
3. I understand completely why people believe there is a god — primarily for comfort that they are being protected and secondarily because this theory provides explanations for unknown phenomena.

(Age 14)

31 July 1971

But is He useful?

I have never believed in God or immortality, but always vaguely sensed that such beliefs were good, on the whole, for a society — because they tend to be accompanied by moral codes that ensure practices valuable from society's point of view. I therefore find Bertrand Russell's distaste for all of religion very interesting. He argues that one should never praise an erroneous belief for its social aspects; that an attempt to prolong beliefs which are unsupportable for the social good they may do is bound to result in social stagnation, repression of analysis, and other evils associated with dogmatism.

(Age 15)

4 March 1975

An explanation of haloes

Attending Sunday morning service in Northampton this weekend, I was struck by an optical illusion. At the pulpit a woman was reading from the Bible and *Science and Health*. I looked at her as she read, my eyes fixed on her face. Behind her head was a blank white wall. My retinas grew accustomed to the dark shape of the head against the wall, so that as she swayed slightly with the reading the field of my retina bleached by the brightness shifted around, throwing bright light on relatively un-bleached cells along the outline of the head. The result: I saw a shimmering halo around the reader's head.

I think it would be stretching it to suppose that this illusion is responsible for the appearance of haloes in religious traditions.

(Age 19)

8 November 1980

Resolution of Pascal's wager

Pascal's wager runs like this: most of us believe there is a positive net probability, even if it's microscopic, that leading the life of Christian virtue will earn us an eternity of bliss. Since this pious effort is finite and the expected reward infinite, obviously we would be fools not to devote our lives to virtue.

It is not enough to ridicule these crazy old arguments; we must either show they are nonsense or refute the logic. I don't think this one is nonsense.

My father and Carolyn incline to the resolution that the probability is not positive but in fact zero. But I don't think that is true for most of us; and it certainly was not true for the average skeptic of earlier centuries, when a relatively uniform Christianity prevailed.

Here is a different possibility. In what sense must a pleasure next year be worth the same to me now as the same pleasure immediately? After all, a pleasure given to my neighbour is not worth the same as a pleasure given to me. Isn't my future self also a slightly different person from me-now? Suppose we persuade ourselves that me-after-t-years is only the same as me-now up to a factor $\exp(-\alpha t)$ for some small positive α. Then no matter how small α is, the expected reward of virtuous behaviour is rendered finite.

8 June 1985

Bluejays, gargoyles, and creationism

Lee found a feather from a bluejay the other day, and pointed out that only about 15% of it was actually blue. Most of one side and all of the other — the parts you don't see when the wings are folded — was grey.

This reminded me of the old story that when cathedrals were built in the middle ages, the craftsmen made it a point of pride to carve gargoyles and other high-up statuary not just on the side facing the people below, but on the back, too — because *God* can see everything.

There's a discrepancy here. Why should we take pains with the flip side of a gargoyle for a God who cut corners on the bluejay?

20 June 1985

God as a scientist, or farmer

As science advances, the temptation to assume the existence of a God to explain the natural world becomes less compelling. Ironically, however, our growing scientific abilities make it easier to see how a God might exist after all — in the form of a scientist (or farmer?) in some larger universe who has created us for some purpose of his own. I would say that the probability that what we think of is the universe is contained in some larger being's laboratory is — oh, 10^{-6} or so. That's a higher estimate than I would give for the existence of a Creator in the more traditional sense, the sort endowed with perfect goodness, absolute wisdom, and other such properties that cannot be defined and thus remove the subject from rational discussion.

The God of the religions is eternal, and his origin admits no explanation — which is nonsense and gets you nowhere. My God, in contrast, probably evolved by natural selection.

8 October 1985

Belief in God

If I had been brought up believing in the existence of God, would I have had the independence of mind later on to discard that belief? For years I was troubled by this question, but lately, having accumulated more experience of people, I have decided that the answer is probably yes. In fact I have identified one or two atheists in my acquainttance who would *not* have been atheists, I suspect, had they had religious upbringings. But I believe my habits of thought are more analytical than theirs.

This view of my own atheism makes the prevalence of religious belief among the philosophers of the past harder to explain. It would be one thing if the great philosophers had held an erroneous belief because they never analyzed their faith, but this is not the case at all. From Aquinas to Descartes to Spinoza to Kant, the question of God has been analyzed with care and concentration. Yet one after another, these men produced proofs of His existence incredible in their variety, but all deciding in the affirmative.

Since I do not claim to be more astute than Descartes, how can it be that I see more clearly? I think the explanation lies with the progress of science. The question of God is easier now than it was then.

31 August 1986

Irreligious intolerance

The world is full of Christians and Jews, Hindus and Muslims, and between religions the rule is intolerance. On the other hand a good atheist lumps all religions together as delusions, the distinctions between them unimportant.

It is a cruel trick of circumstances that from this point of view of intolerance, atheism cannot clearly be distinguished from a religion. Hindus don't think the distinction between Christians and Jews is very important, either. I believe that lack of religion is fundamentally different from religion, but I would hardly care to argue the point with someone who disagreed.

For despite its different philosophical status, atheism functions psychologically much like any other religious affiliation. I see this in myself. When I contemplate a woman I've met as a potential wife, few qualities are more damning than religious belief. You might think my impiety should be more offensive to her than her piety to me. But I am more serious about my atheism than most Christians are about their Christianity! Anyone who is religious has sealed off a certain mental territory from rational analysis, and that, to me, is frightening.

21 September 1990

Ontological grook

As God exists, so does the Devil,
And here's a proof that's rather terse:

Existence makes a good thing better;
It also makes a bad thing worse.

12 January 1993

Existence of God; existence of secretaries

Centuries ago, there was no doubt as to the most compelling evidence for the existence of God: it was the world around us. Today the world has been largely accounted for by other explanations, but God has not gone away. He lives on, justified by a collection of subtler arguments that used to be of secondary importance but have expanded to fill the void.

So it is with secretaries and typing. For centuries, the central function of secretaries has been document preparation, but nowadays, in certain circles at least, we and our computers have largely taken that function from them. Yet the secretaries have not gone away. They live on, justified by a collection of other functions that used to be marginal but have somehow moved to the center.

Long-term, I think both God and secretaries are in trouble.

27 January 1995

Might God be a woman?

Emma asked the other day, "Who made the world?" I found myself answering, "Nobody. But many people believe that a man called God made the world."

I intentionally said "man" rather than "man or woman" or "person" or "being". My reasoning was as follows. I was describing a belief held by a billion or two people that I myself consider fallacious. However much it may jar with our own notions of how power should be shared between the sexes, it is a fact that the great majority of those billions still conceive of God in the old-fashioned way — as distinctly male.

Yet how uncomfortable I felt to have said "man"! Such sexism! And, sure enough, when I mentioned the occurrence at a dinner party, two female acquaintances expressed shock and disapproval that I had told Emma such a thing.

A gentleman would forbear to describe the details of a particularly ugly murder, though of course he himself had no part in it. It would seem that in much the same way, certain philosophical views, even though held by others one considers misguided, may be too repugnant to be mentioned.

20 August 2004

Religion as carbon monoxide

Driving across the US this summer, we listened to the radio. It was frightening. In large swathes of the west, all you hear is evangelical Christianity. It is common to cruise the dial and find, say, four stations, three of which are running Christian talk shows. Part of what these shows are up to is the usual posturing against liberal targets like gun control and working mothers. The bigger part is just endless, mind-numbing chatter on the theme of, have you truly accepted Jesus?

Millions of Americans must give their attention to this empty stuff. When going on about Jesus, I wonder, are they in fact actually talking about anything at all? You might think it is harmless enough, until you reflect on how such talk must consume people's time. Jesus must be the *only* issue of a public nature that millions of Americans think about in a sustained way. Matters of substance, in particular the paying attention to the world outside the USA, get squeezed out. The result is George W. Bush's catastrophic presidency of wars and simplifications.

Jesus is poison to your brain as carbon monoxide is poison to your blood, hijacking your haemoglobin so it cannot do its work. Further east, there is a poison called Allah.

26 December 2004

Jesus Christ and the gold standard

On Christmas Eve I was listening to a choir service on the radio and thinking, how I love this tradition! But what will happen to it in the decades ahead, if belief in Christ wanes? It would be terrible to lose Christmas!

I think we needn't worry. Economists used to think that paper currency had to be backed up by the promise of repayment in gold. Then they took the gold away, so the paper stood for nothing at all, and lo, it didn't make much difference.

20 July 1971

Evil

For some time I have been uncomfortable with the word "evil". I think it is too melodramatic, and not really applicable to human affairs. I have the feeling, when evil is mentioned, that someone is putting on airs, talking about something high and mighty that doesn't really exist.

(Age 15)

24 November 1974

The evil of murder, the beauty of woman

Against the point of view that there is a transcendent "ought" which makes some actions morally right and others morally wrong, I think a comparison between the evil of murder and the beauty of woman has force. Many would say that murder is intrinsically evil, apart from any negative consequences which are likely to follow for the murderer; that the general human belief in the evil of murder is more than an instinctive and/or cultural piece of baggage. Not so many would argue that the female form is intrinsically beautiful, beyond the earthly desires of human beings.

If you agree that esthetics is not properly a philosophical subject, two options are available: throw out ethics along with it, or justify exalting one and not the other. I see no such justification.

(Age 19)

15 July 1976

Moral imperative as infinite regress

The notion of moral obligation may reflect a real psychological state, but it has no content whatsoever as an intellectual concept. The essence of the moral imperative is that something must be done for its own sake, for no reason. It is impossible to put any meaning into this disembodied "must".

Indeed, the only sort of meaning there could possibly be in saying that one "must" do A is that A is a means to achieve some morally desirable end B. But why then must one pursue B? The prospect of this infinite regress shows that any attempt to give intellectual meaning to moral obligation is as doomed as the First Cause argument for the existence of God.

17 July 1976

Reward in heaven

The moral imperative by its nature is an unsupported "must", but Christianity does not preach it in this pure form. Quite the contrary, the whole structure of Christian metaphysics seems to stand on duty to support moral law: an eternity of bliss and divine favour rewards good conduct, and destruction punishes bad. Heaven and hell are more than backing enough to ensure that good behavior in this life is squarely in one's self-interest. So are the more modest psychological consequences that a less orthodox moralist might adduce to support his moral code.

Either way, the religion is playing false with you. The prescriptive spirit of Christianity is moral, not rational, and there's no compromising the principle that you must act properly because it's right, not because you'll be rewarded. Heaven is an idea that had to be invented because without it the code too obviously has no foundation, but its character is foreign to the fiber of Christianity. I think that is why heaven is such an anaemic concept in Christian tradition.

6 February 1981

Morals cannot rule on little things

My moral sense doesn't merely whisper to me that some actions are good and others not good, but more — that even in the smallest matters there is always one morally best course of action. Must I work another hour, or may I read *Newsweek*? I have an absurd feeling that there is a correct answer, that if I could only consider the issue carefully enough, it would be clear that one action was better than the other in an absolute sense.

This is nonsense, although such feelings have the good effect of making us weigh our actions more carefully. Moral principles have no fundamental reality. They are not physical laws with a chance of being exactly *true*: they are approximations that function only in the macroscopic world of human affairs. (Of course we can devise models of morality that can be precisely defined and applied, but such models invariably leave out part of the real sense of what is right and wrong.) Hence they cannot bear close reasoning without generating inconsistencies, which means they cannot judge delicate questions.

26 January 1985

Subway ethics

This evening I ran into an ethical dilemma of the smallest possible size. I arrived at the Central Square subway station around midnight with a token in my pocket, but since the train wasn't there, I decided I'd buy some new tokens anyway to save time tomorrow. But when I offered the man in the booth my five-dollar bill, he pointed at a locked cash box to indicate he had no change, and waved me through the turnstile for free.

Now should I have revealed that I had a token after all, and used it?

What appeals to me about this question is that although it sounds reasonable, it isn't. Because of its diminutive scale, there are those who might consider it an interesting test of the fine points of this or that theory of ethics. But such "shoulds" have none of the meaning people pretend they have, and this question has no answer.

26 July 1991

Can an adult sin?

One of my mother's favourite stories from my childhood goes like this. One day we were discussing sin. To my surprise I discovered that this was something grownups worried about too. "But what sins can *grownups* have?" I asked.

Of course the joke is that grownups' sins are giant-sized compared with children's. Yet in a deeper sense, I was right. The essence of sin is action that contradicts principles laid down by a higher authority. For children, the higher authority is their parents. For grownups, what is it? Three conventional answers are (1) God, (2) some sort of absolute moral truths, and (3) society (or, perhaps, our genetic blueprint). The awkward thing is, the first two of these candidates do not actually exist, and most people consider the third insufficient to give morals their zip. So the question of whether grownups can sin is not so silly after all.

12 January 2002

Responsibility decays with distance

We care more about poverty here than in Afghanistan (per capita), more about a murder here than in Burundi. Many people feel that this unevenness in our attentions is unjust. They say, is an Englishman more human than an Afghan? They tend to feel guilty about their own unevenly spread sympathies or actions, or disparaging of others'.

I think that feelings naturally decay with distance, whether physical or psychological, and that although one should strive for the decay function to be not too rapid, one should not feel guilty that it is there at all. Here's a thought experiment. Suppose the world were unbounded, populated by infinitely many people living in lands receding infinitely into the distance. Should we care equally about all of them? Of course not, for this would mean caring zero about each one. But now, surely our view of the world cannot depend on the fact that 6 billion is less than infinity.

The principle generalizes. It is natural, and not reprehensible, that we care less about chimpanzees than humans and less about squirrels than chimpanzees. And it is natural that we care less about ourselves ten years from now than ourselves tomorrow. Feelings decline with temporal distance as well as spatial, and that, as I've argued elsewhere, is the proper resolution of Pascal's wager.

18 August 2002

Wickedness and destruction

It is an old idea that man's wickedness will bring destruction upon him. As the scenarios of doom have multiplied in the last century, this prophecy has taken new forms. Capitalist greed will destroy the environment and we will be poisoned, or starved. Our hubristic playing at God in the test tube will unbalance the human gene pool. Licentious enjoyment of sex and drugs will unleash superdiseases that bring civilization crashing down. Or the military men in their obscene lust for power will trigger a nuclear war. Take your pick.

These apocalypses may be plausible technically, but they are inflated morally. Man in his arrogance is determined to believe that if a big destruction comes, it must be the result of a big wickedness. I don't think so. Our wickedness hasn't changed much since Biblical times. If we destroy ourselves, it will not be because our souls have become ever darker, but because our technology has become ever stronger. Our death, like most deaths, will not be retribution; it will be meaningless.

3 December 2005

Tugging on an asteroid

Lu and Love published an intriguing paper in *Nature* last week. Suppose an asteroid is heading for Earth and may hit us in a few decades. How can we prevent a calamity? Well, the idea is that perhaps we can fly a rocket near to the asteroid and have it hover on one side for a long time, as long as a year. Its little gravitational tug, added up over the whole year, might change the asteroid's orbit enough to save us!

Impossible, you say. The asteroid is vast and the spacecraft just a few tons. How could its minuscule little gravitational pull make the difference? But the answer is in the equations. A minuscule tug applied gently and persistently for a very long time can nudge even a planetesimal to a new trajectory.

And now I imagine this as a moral for life. How can a decent ordinary person have any effect amid the swirling vast forces of armies and economies and multinational corporations and George W. Bush? Well maybe, just maybe, if you take up a position on the right side of the crowd and stick to it persistently for a long time, you may change the world.

4 December 2005

300,000 graduates a year

We watch the news and see suicide bombers, scandal and corruption, war and genocide. A wicked person can do so much damage! — so much more damage than a good person can normally do good. How can good beat evil in the face of such asymmetry?

Through sheer numbers. I remind myself that every year in the UK, 300,000 new university graduates are produced. A thousand a day! That's a Mississippi River of mostly decent and ordinary educated people flowing into the system, year in, year out. The wicked ones spread their turbulence, but it's reasonable to hope that they can't change the main course.

Science

23 November 1970

The future is determined

Everything that happens happens as the direct result of the laws of nature acting upon molecules, photons, electrons and so on. The future is completely determined at this present instant, for all actions must result directly from the laws of nature; there can be no exceptions unless one believes in a god. Every thought I will ever think is right now "in the cards" and I have *no* control over what I will do and be in the future. Everything is planned. If I decide to work harder I will work harder, of course, but the plain fact is that whether the molecules of my brain decide to work harder or not is something that is determined completely and without a doubt by physical laws.

(Age 15)

16 June 1973

Sensitivity to initial conditions

I am frequently annoyed by discussions of the uncertainty principle that seem to muddle through the fundamental qualitative difference between large predictability and complete predictability. For example, saying that brain processes are sufficiently macroscopic as to be practically unaffected by molecular uncertainties seems to me a misguided argument; the issue is not whether events are "practically" determined, but whether they are determined.

However, out of the muddle comes a question that has interested and baffled me: how rapidly do little events in the world get amplified? That is, if I could magically change the position of one water molecule in the ocean by 6.4 angstrom units right now, how different would the world be in an hour or ten thousand years from now? Is the pattern of events so delicate that such a change would add or subtract three deaths to India's total in 1978? Or would the change do little more than shift the positions of a few neighboring molecules?

(Age 17)

11 September 1980

Ontogeny recapitulates phylogeny

I remember that as a child of ten or eleven I had a lively instinct to theorize and classify. For example, I thought it would be interesting to conduct a study of the various forms of fire. Fire could be made by burning wood or paper, by igniting hydrogen and oxygen, and also, I had read, by mixing potassium with water. This raised the question, how many different kinds of fire are there altogether? It seemed to me that it would be a very worthwhile project for someone to compile the complete list.

The truth about fire and other things, I later realized, involves a bit more mechanism and a bit less taxonomy. I find it remarkable, though, how my own early development parallels that of early scientists in history. In my enthusiasm to classify and make lists I was repeating a tradition of generations of alchemists and philosophers from Aristotle to Bacon. At this stage of intellectual activity the passion for learning is genuine, I believe, and what's missing are only the skills of the trade.

11 August 1989

A principle of good science

If the answer is highly sensitive to perturbations, you've probably asked the wrong question.

29 December 1990

Fractal structure of scientific revolutions

Thomas Kuhn became famous by arguing that the history of science is one of stable periods of "normal science" punctuated by sudden "paradigm shifts". At these revolutionary moments it is not just theories that change, but whole perceptions of the world.

The trouble with this view is that it is impossible to make an absolute distinction between what is normal and what is revolutionary. Therefore I prefer to picture the course of science as a *fractal*. Progress occurs by jumps, yes, but they are jumps of all sizes, with paradigm shifts to match. What's revolutionary on one scale is normal on another. This conception of science is more prosaic than Kuhn's, indeed fundamentally more old-fashioned, but I believe it is a truer one.

Both Kuhn's theory and (more recently) fractals are so famous that I am sure the idea of combining the two is not original. In fact if fractals had been in the air at the time of Kuhn's work, he would probably have thought about them in this context himself. Would he still have dared to write the book he did?

11 June 1991

Hubris and meta-hubris

Until Einstein and Heisenberg, so the well-known story goes, scientists assumed there were a finite number of laws of physics and that one day we might find them. This confidence was destroyed by the arrival of relativity and quantum mechanics.

The response to this upheaval has been a more or less universal conversion to a different conception of science. It seems confidently accepted in all quarters now that the hope for a complete understanding of the laws of physics was naive. Science is now viewed as an endless series of approximations — and not necessarily approximations *to* anything. How childish we were to think otherwise!

In my opinion the new attitude is no better justified than the old one was. After two centuries of Newton, we extrapolated cavalierly to a future of perfect understanding. After one century of upheavals, we have extrapolated just as cavalierly to a future of unending upheavals! We believe we have figured out the secret — that there is no secret. The best you can say for this view is that it is a subtler form of hubris.

3 July 1993

Genuine data

At the Householder meeting the week before last, we accumulated a graph of people present (names written in circles) and relationships of coauthorship between them (lines connecting those circles). As names and links were added, John Gilbert and others typed them into the computer for subsequent analysis of connectivity, planarity, cliques, small separators, etc.

Typing in the data started out fun, but soon became a headache. It was fun when there were a dozen names and forty links. It was a headache when there were more than a hundred names and five hundred links. And at about that time I noticed — errors were creeping in! Missing links. Erroneous links. Misspellings. It became evident that this project had grown to the point where we could not with a reasonable effort expect to make the dataset entirely correct.

What had started out a game had become real.

This experience suggests a principle that I wouldn't care to defend too seriously. A genuine dataset is one that contains errors.

3 May 1996

My father's definition of a boat

My father has been investigating ideas related to sailboats. The essence of the matter is that there are two foils, connected somehow, and that they are immersed in fluids travelling at different velocities. Given that, anything is possible — including, most notably, motion upwind.

We all know the line about a chicken being an egg's way of making another egg. My father's thoughts on sailing suggest an analogous definition for that subject area:

A boat is a device for connecting a sail to a keel.

20 January 1997

Large and small parameters

If no parameters in the world were very large or very small, science would reduce to an exhaustive list of everything.

Stars and Planets

21 February 1981

Stars and constellations

Until a month ago I recognized nothing in the sky at night but the moon and the dippers. Everything else was randomness, and I paid no attention to it. Suddenly Carolyn and I bought a few charts and books, learned the names of ten constellations and twenty stars, and the night has changed for me completely. I see something. I cannot keep my eyes away from it. I am reading books on astronomy. I have gained a new friend.

It is hard to imagine a clearer illustration that the mind needs to be stimulated by a little outside information. Twenty-five years under the stars didn't teach me as much as the last ten evenings. With the right help in the right environment, we pick up all kinds of knowledge effortlessly. Thank goodness my parents were not as remiss in most fields as they were in stargazing.

29 November 1981

Planetarium illusion

We visited the planetarium in Golden Gate Park today. After we had looked at a firmament of thousands of stars for a while, the whole sky began to rotate slowly. The sensation was overwhelming that they were fixed and *we* were moving.

What I wonder is, was this a purely "optical" illusion? Would the perception of motion have been as strong if the pattern projected on the roof had been abstract? Or did our ingrained knowledge that the stars are fixed contribute to the effect?

Both?

7 December 1981

Constellations in the coffee shop

I walked up to Marcel's coffee and pastry counter in Tresidder Union and noticed that the thin dark girl who works afternoons was at the register, but the plump early mornings girl was also still visible in the back. Only at the second blink did I realize that I had perceived them both as stars in the sky — just as in the night in recent weeks we see the winter star Capella rising in the east, even while the summer star Vega is still visible in the west.

This is an ordinary enough analogy. What impresses me is that it was no second or third thought, but the very first thing I sensed as I entered the building. The status of analogies as a tool of reasoned analysis is shaky, for they are contrived and dangerous; but there can be no doubt of their fundamental status in the brain. We build our concepts out of primitive analogies, more than the other way around.

1 August 1984

Extraterrestrial life and the speed of light

I consider the lack of evidence of other life in the galaxy a scientific datum of the first order. The essence of life is to expand into new territory, almost by definition: the organism that does this ends up observable. One would expect the galaxy to have numerous civilizations vastly older than ours. Where are they?

I have heard various theories to explain their absence, such as the idea that civilizations invariably self-destruct at some stage of their development, or that the aliens *are* here but we are too primitive to detect them. But I find these explanations unconvincing.

My best guess is that the galaxy is simply not old enough for life to have spread through it. As a rough estimate, suppose that one star in 1000 has originated a civilization 10^9 years old, and let a typical distance between stars be 10 light years. Then if we have observed no alien life, an upper bound on the rate of expansion of civilizations is around 100 light years per 10^9 years, or 60 miles per hour.

Curiously, this apparent speed limit gives an indirect argument that no escape will ever be found from Einstein's theories — for what could possibly hold the expansion of life to 60 mph if space warps and hyperdrives were possible?

5 May 1985

The fluid earth

If the earth's mantel were not in some respects a liquid, there would be no convection in it and no continental drift. If the earth's core were not liquid, there would be no flow of the iron there and no resulting magnetic field. It is the earth's magnetic field that has led to our discovery of continental drift. So what one geophysical fluid process has caused, another has revealed.

21 December 1990

How far is Perth from Boston?

Perth, my father told me years ago, is about as far as you can get from Boston. Visiting Australia has made me curious to pin it down precisely. Just how nearly antipodal are they?

(a) If Boston were the North Pole, what would be the latitude of Perth?
(b) What proportion of the earth's surface is further from Boston than Perth?

Answering such questions proves surprisingly tricky. One approach is to convert from polar coordinates to x, y, z, then take an inner product. If the latitudes and longitudes are θ_1, θ_2 and φ_1, φ_2, you end up with the following formulas:

(a) $(1+\cos\alpha)/2$, (b) α,

where $\cos\alpha = \cos\theta_1 \cos\theta_2 \cos(\varphi_1 - \varphi_2) + \sin\theta_1 \sin\theta_2$.

After consulting an atlas we conclude: if Boston were the North Pole, Perth would be in the Antarctic at 168° 14′ S, with only 1.1% of the earth further away.

27 July 1991

Life on planets of other suns

There is evidence this week that a certain nearby pulsar has a planet orbiting around it. If so, this may count as the first discovery of a planet outside our solar system. Big news! Anne even heard it on television. For if there are planets out there, the newscaster pointed out, this increases the probability of extraterrestrial life!

Does it? I propose that the opposite conclusion is also reasonable: that the probability of extraterrestrial life went down this week.

My reasoning turns on the fundamental mystery of extraterrestrial life: if they are out there, why don't we see them? If planets are rare, then the mystery is not so hard to explain: other civilizations just haven't had enough time, at whatever speed galactic civilizations tend to expand, to reach us. The more common planets are, the more strained this explanation becomes and the easier it becomes to conceive that life on Earth may be unique.

1 January 2000

Millennium fireworks

There were spectacular fireworks around the world last night as humanity enjoyed its biggest party ever. The celebrations rolled around the globe for twenty-four hours, from Auckland to Sydney to Hong Kong, from London to Rio to New York, from Houston to Los Angeles to Honolulu — always at midnight local time.

It might have been interesting to an observer out in space. In the middle of the dark side of the earth, at an arbitrary moment, an arc of colorful lights suddenly switch on, sparkling irregularly from pole to pole. The sparkles continue in place as the earth rotates underneath, one full turn. Then, just as arbitrarily, they switch off.

13 October 2003

The other moon illusion

The famous optical illusion associated with the moon concerns its size on the horizon and up in the sky. Lately I've noticed another illusion concerning its shape, and this one is powerful too.

Look up one day when both the moon and the sun are in the sky. The moon will be shaped like a C or a D, with the circle of its edge on one side and the straighter shadow line on the other. Now look at the sun again and try to understand why that shadow line is where it is. *It's completely wrong!* Your eye tells you, *the moon should look fuller!*

Like the more famous moon illusion, I think this one is caused by confusion about distances. You, the moon and the sun form a triangle in which the sun is effectively infinitely further away than the moon. However, your eye cannot perceive this. It interprets the moon and the sun as bodies at comparable distances. This makes the trigonometry all wrong and the position of that shadow inexplicable.

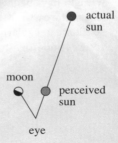

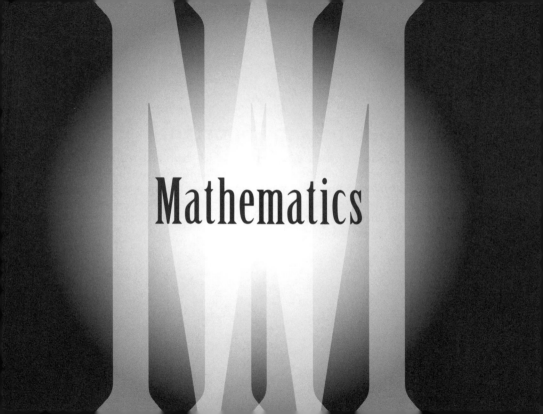

14 October 1973

Why are numbers so important?

In my view mathematics is that area of thought which deals with the logical manipulation of homecooked definitions. Isn't it remarkable how much numbers are entwined in the subject! Some areas of mathematics use numbers less than others, but as far as I can see at this point almost all rely on them heavily, either directly or indirectly. Why does this single notion of quantity prove so valuable to so many forms of complex logical thought?

(Age 18)

20 October 1977

The sparseness of mathematical concepts

One might imagine that as one climbed from lower to higher levels of mathematics new definitions would grow rarer and theorems based on old definitions would come to predominate. But we don't observe this at all — mathematics at all levels demands a steady supply of new definitions. There's a certain sparseness in the interrelationships among mathematical concepts.

27 June 1980

What numerical analysis is about

Some people mistakenly believe that numerical analysis is the study of rounding errors on computers. In fact, thank goodness, it is closer to the truth to say that it is the study of efficient algorithms (for mathematical problems). If computers had infinite precision numerical analysis would still be here; if they had infinite speed, most of it would not.

2 July 1983

Knowing what we're talking about

Often mathematicians make slow progress or even reach false conclusions because they launch too quickly into high-level abstractions that they cannot grasp very fully. One wishes they would spend more time with concrete examples — learn what they are talking about!

Yet the power of mathematics lies ultimately in the fact that one does *not* know what one is talking about. If you fully comprehend the meaning of your theorem, it is a dull tool that will apply to nothing unforeseen. The trick is to use concrete examples like booster rockets, travel as high as possible on them but then let go.

17 May 1987

Money and numbers, apples and oranges

Numbers are to computation as money is to commerce. Without money, commercial goods would occupy a space of bewilderingly many dimensions; we would have to trade apples for oranges. In one stroke money makes everything comparable on a one-dimensional scale.

Similarly, scientific computations should in principle deal in symbols representing a boundless collection of physical quantities. Computer programs like Macsyma, SMP, Maple and Reduce have been written to make it possible to carry out just such computations mechanically. But this kind of symbolic computation is hard, and simple problems can easily generate vast expressions that exhaust a computer's memory. It is a shame to narrow one's vision by replacing those symbols by mere numbers, but when one is able to do so, what a saving!

30 November 1988

Digits of π as a function of time

The plot below, based on figures given in the marvelous book *Pi and the AGM* by Borwein and Borwein, indicates roughly how many digits of π have been known as a function of time since 1600. The scale is logarithmic.

Can you spot the invention of computers?

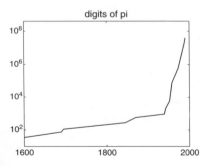

20 March 1989

Mathematical Engineering

How about a field called mathematical engineering? We already have mechanical, civil, chemical, nuclear, aeronautical, and electrical engineering, and in most cases, naturally enough, the phrase denotes the practical, technological end of a scientific discipline. Why not "mathematical engineering" for the practical end of mathematics? These days, of course, a great deal of the subject matter would be computational.

I'm about half serious. Mathematical engineering is not such a euphonious phrase that I'll campaign hard for it, and in that sense I'm not so serious. However, there's a serious reason why I regret that we don't use some such expression. Nobody seems confused about what electrical or chemical engineering is good for, but I've found it quite hard to explain numerical analysis. People lack a pigeonhole to put the field in. The phrase "mathematical engineering" would make it plain that it's perfectly natural for mathematics, like other fields, to have its nuts and bolts practitioners.

16 January 1997

The spectrum from pure to applied

No matter where you sit on the pure-applied spectrum, your perception will be the same. Your own position seems the middle one, with equal expanses of more applied work stretching off in one direction and purer work in the other.

3 February 1997

Discrete and continuous

The big gulf in the mathematical sciences is between the continuous problems (and people) and the discrete ones. Most scientists and engineers are in the continuous group, and most computer scientists are in the discrete one.

5 February 1997

Two unsolved problems of mathematics

The two most important unsolved problems in mathematics are the Riemann hypothesis and "P = NP?". Of the two, it is the latter whose solution will have the greater impact.

13 February 1997

The obscurity of numerical analysis

As technology advances, the ingenious ideas that make progress possible vanish into the inner workings of our machines, where only experts may be aware of their existence. Numerical algorithms, being exceptionally uninteresting and incomprehensible to the public, vanish exceptionally fast.

14 February 1997

Unsolved problem of numerical analysis

Is there an $O(n^{2+\varepsilon})$ algorithm for solving an $n \times n$ system $Ax = b$? This is the biggest unsolved problem in numerical analysis, but nobody is working on it.

1 March 1999

Rigor in 20th century mathematics

One might say that the greatest drama of the mathematics of the first half of this century was the development of modern analysis, beginning with Cantor and Lebesgue and leading to the powerful methods of dealing with limits and infinities and functions that we now take for granted. And for the second half of the century, the biggest story has been nonlinear dynamics, from solitons to chaos to cellular automata to inertial manifolds, all of it unleashed by the computer.

This bit of history may shed some light on prevailing views about mathematical style. Mathematicians believe in rigor, and to applied mathematicians like me, they sometimes seem to carry this belief to an obsessive extreme. They don't care about phenomena, it sometimes seems, only about technique. But perhaps history explains this style. The very essence of that early-century triumph was technique; the ideas of measure theory make little sense except expressed as theorems. Nonlinear dynamics is a different kind of field, more like a science of the usual sort, whose natural state is a flourishing mix of theory and experiment. But cultural attitudes take a while to catch up.

5 February 2001

Marx and mathematics

Marx's labor theory of value held that the value of an object is equal to the amount of labor put into it, no more and no less. If the price of a good doesn't match the labor in it, then somebody is making an immoral profit.

Like most flawed ideas, this one correlates with the truth, but that's the best you can say for it. Economists long ago realized that the labor theory doesn't do much to explain why gold costs more than iron, why J. K. Rowling is a millionaire, or why corn grown on rich soil in Indiana is worth as much as corn grown in the rocks of New Hampshire. You can find immoral exploitation in each case if you look hard enough, but you'll be seeing the world askew.

Mathematicians are in the grip of a similar fallacy. They like to judge a mathematical contribution by how *difficult* it was. Profound innovations get little credit if it didn't take a genius to originate them. What a simplistic view of the world! Of course major breakthroughs often develop through genius, for somebody has to come first, and the geniuses, by and large, are ahead of the pack. But that's a matter of statistics, not necessity.

4 December 2001

How to think unrigorously

E., a bright but all too pure Junior Research Fellow in mathematics, caught up with me on the St Giles' pavement one afternoon. There's something I've been wanting to ask you, she said earnestly, because I just don't understand. When you investigate a mathematical problem, Nick, if you don't go about it rigorously, how *do* you go about it?

A few yards further along I ran into my old applied mathematical friend Andrew Fowler and reported to him E.'s question. In his dry northern Irish accent he instantly rejoined, "Yes, well, and how does she get out of bed in the morning!"

11 August 2003

What does it mean to solve a problem?

At ANU recently I had lunch with some mathematicians. One of them* pointed out how profoundly our views have changed with the years. Traditionally, he said, to "solve" a mathematical problem meant to prove existence of a solution. Modern developments, however, have transformed our thinking, and now it is recognized than an equally valid idea may be that to "solve" a problem is to elucidate general geometric and topological features of solutions of problems of that class.

There was no mention of a third idea. One might say that to solve a problem means to actually calculate the answer. I wonder whether this notion will catch on?

*Terence Tao

19 May 2006

Mathematics and fracture mechanics

Are there examples of great errors in mathematics, of big "theorems" that were believed for years until an error in the proof was found? I can't think of any. Certainly they aren't common. Yet all of us in the business know that the literature is full of mistakes on a smaller scale. Little papers with little theorems of little impact often "prove" them falsely. We've all seen this.

So does this mean that the mathematicians who prove bigger theorems don't make mistakes? No, I think the explanation is different: the errors in big results almost instantly get corrected, before they do any damage. This is just like the physics of the fracture of solids. Any piece of glass is full of cracks and fissures on a small scale, small enough that ordinary forces on the glass don't make them grow. These linger unchanged and do no harm. Above a critical size of crack, however, the balance tips and stresses suddenly cause unstable growth of cracks: fracture and failure.

And just as glass is strong despite those little cracks, so mathematics is strong despite those little errors.

The same analogy applies to bugs in computer programs.

9 November 1976

Annihilating 50 million people

The other evening it emerged that Greg thought Russia had a population of 250 million people, while I thought it had more than 300 million. We challenged each other, looked it up in the almanac, and found Greg was the winner.

I was struck at first with the strangeness of the thought that I could so carelessly conceptualize fifty million warm bodies in and out of existence. But then, I thought, in what sense can I conceive of 250 million or 300 million people anyway? Three people, I can visualize in my head. Thirty thousand is a baseball-game full. But three hundred million? The gross population of Russia exists only mathematically. I can compare it to the American population by computing a ratio — so Greg's answer did have some meaning to me — but by itself the number has scarcely any substance whatever.

6 April 1984

Explaining my position in humanity

There are some 10^{10} people in the world. Roughly speaking, I am among the top 10^5 most successful. Of course the precise ranking depends on exactly what combination of abilities, fame, influence, wealth, etc. you measure.

Accounting for this improbably high position in the hierarchy is really a troubling problem for me. I don't actually believe this, for among other things it seems implausible dimensionally, but a tantalizing idea suggests itself: can one argue somehow that in a random population of n people, the expected position to find an introspective person of my sort is on the order of $\sqrt{n}$ from the top?

2 August 1985

Ten billion

Roughly, the brain contains 10^{10} neurons and the genetic code contains 10^{10} nucleotides.

As a rule of thumb, perhaps 10^{10} components is the level of complexity at which a system may go so far beyond the behaviour of its individual pieces as to exhibit a life of its own.

The earth has around 10^{10} human inhabitants. Perhaps the truest understanding of them comes from ignoring the individual particles and viewing humanity instead as a single organism, occupying, and manipulating, the whole surface of the earth.

24 March 1988

Beyond Roman numerals

The Romans used Roman numerals: VIII.

Then came decimal representations, logarithmically more compact: 5111.

Then came scientific notation and another logarithmic compression: 5×10^{111}.

Will history stop here? Of course mathematicians have iterated the process of taking logarithms, but that is quite a different thing from the common practice. Will we one day habitually deal with numbers so vast that a triply logarithmic notation becomes standard?

6 June 1996

Music and megabytes

Computer science is built on terms like kilobyte, megabyte, gigabyte. This system of terminology takes advantage of a convenient coincidence: 2^{10} is approximately equal to 10^3. "Kilo", in computer science, really means 1024, but that exceeds 1000 by only 2.4%, and analogously with "mega" and the rest. Prefixes representing 10^2 or 10^5, say, would never have worked so well.

This is analogous to another coincidence, the one that gave us the 12-tone scale in Western music. By a stroke of luck, $2^{3/12} \approx 1.1892$, $2^{5/12} \approx 1.3348$, and $2^{7/12} \approx 1.4983$. These numbers are strangely close to 5/4, 4/3, and 3/2; the errors are just 0.90%, 0.11%, and 0.11%, respectively. Thanks to these close agreements, chords played on an equal-tempered scale sound harmonious. Dividing the scale into 11 or 13 or some other number of tones would not have worked at all.

19 March 1997

One way to put together 10^{44} pieces

Life on Earth consists of 10^7 species, each consisting of 10^{10} individuals, each consisting of 10^{13} cells, each consisting of 10^{14} molecules.

17 September 1999

Molecules and floating point numbers

Our everyday world is made of objects that typically contain, say, 10^{30} molecules. Such a number is so large that these objects behave like continua. Thanks solely to the statistics of large numbers, they obey precise physical laws, like the gas law that says that if the air in this room is made 1% hotter, it becomes 1% less dense.

Since $10^{30} = (10^{10})^3$, we may put it another way. A human-scale object contains 10^{10} molecules from one end to the other — close enough to infinity for most purposes.

Now consider numerical computations on computers. We work with floating-point numbers, with resolution about 10^{-16}. This means that if we assign x, y, z computer coordinates to the points in the room, there will be around 10^{16} grid points from one end of the room to the other. Our discretizations are a million times finer than God's.

But do you know something odd? The laws of continuum physics are widely admired as beautiful and reliable, triumphs of the human mind. Yet floating-point arithmetic is widely regarded by the very same people as dangerous in practice and ugly in concept.

6 August 2003

The speed of light is an integer

Emma was reading physics and encountered the fact that in 1983, the meter was redefined so that the speed of light is exactly 299,792,458 meters per second.

Why, she asked, didn't they make it 300,000,000?

Any child of the computer generation can see the answer. It's a matter of backward compatibility, of course.

10 February 2009

Four bugs on a rectangle

Here is the biggest number I have encountered in my work (a joint project with James Lottes and Jon Chapman). You have four bugs initially at the corners of a 2 × 1 rectangle, each walking towards the next at speed 1, spiraling in towards an eventual collision at the midpoint. Consider the distance between two of the bugs on opposite corners. By what factor has this distance decreased when the bugs complete the first of their infinitely many circuits around the midpoint?

The answer is approximately $10^{427907250}$.

Mathematics and Science in Everyday Life

19 January 1986

Ill-conditioning in Leicester Square

Something startled me in Leicester Square last summer, where I was meeting M. and C. for a day on the town in London. There is a pole set in the pavement, and surrounding it, a honeycomb of hexagonal tiles with the names of various cities of the Empire engraved in them, each lying in the appropriate direction.

What startled me was the relationship of Sydney to Auckland. Here was the tile for Sydney, and way over there was Auckland — 40° different! How could this be?

The answer is a matter of what numerical analysts call ill-conditioning, and comes quickly when you think about it. The tiles are placed according to great-circle routes around the earth, and Sydney is nearly antipodal to London. Suppose it were exactly antipodal. Then an infinitesimal perturbation of its location could make the great-circle direction to Sydney lie anywhere on the compass.

Alternatively, notice that the great-circle direction from Leicester Square to Sydney and Auckland are precisely opposite to the great-circle directions to antipodal-Sydney (near the Azores) and antipodal-Auckland (Gibraltar). Since the latter are a thousand miles apart but not so far from London, it's no surprise that there's a big angle between then.

18 December 1986

Why do lights reflect as streaks in a river?

Look across a river at night at the lights on the other side. If the surface is glassy, each light reflects in the water as a point, but more often the surface is irregular and the reflections stretch into long vertical streaks. Why? Why not round blobs, or even horizontal streaks?

The phenomenon is ubiquitous. I see it in the floors of shiny corridors and in headlights on the road in the rain. It's visible in the daytime, too: the image of a building across the river usually appears vertically striped, for the horizontal detail is preserved in the reflection but the vertical detail is lost.

The explanation is just a matter of trigonometry, but tricky. Suppose that you and the object being viewed are at the same height above the water, at a small angle $\theta \ll 1$ from the horizontal as measured from the point of reflection halfway between. Assume that the surface deviates from the horizontal and random by small amounts $\varepsilon \ll 1$ in arbitrary directions. A small deviation in the direction between you and the object shifts the line of sight by an angle ε, while a small deviation in the sideways directions shifts it by only $½\theta\varepsilon$. So the streaks have length-to-width ratio approximately $1/2\theta$.

19 December 1986

Human particles in spatial equilibrium

It is an old marvel of physics that a particle responds instantaneously to the forces exerted on it by so many other particles near and far. How can a mere electron carry out all the required calculations?

I marvel at how I, too, seem unconsciously to perform similar calculations in a crowd of people. Sit me down in a subway car in which every seat is full, and I am quite comfortable, despite the businessman pressed against my left shoulder and the high school girl with the cello on my right. But if the businessman and a number of the other passengers leave the train at Harvard Square, the forces on me are cast into disequilibrium. Immediately I feel a net force pushing me leftward towards a socially more appropriate distance from the high school girl.

Or put me in a crowded men's room at the theatre. With no qualms I will take my place in a row of men standing close-packed at the urinals. But if the two positions to my left become vacant while I'm standing there, suddenly I am indecently close to the fellow on the right. That subconscious force of interaction between human bodies is out of balance again.

18 July 1987

Crockery, cutlery, and conductivity

Different materials conduct heat differently, with plenty of everyday consequences. One is that we put rugs on the bathroom floor — not because wool is warmer than tile, but because it conducts heat away from the feet less efficiently.

This summer, living in England without a dishwasher, I've noticed another domestic consequence of conductivity differences. I wash and rinse the dishes in hot water, then put them on the drainboard. Fifteen minutes later the plates and bowls and glasses are dry, ready to be put on the shelf, but the silverware is invariably clammy and needs a going-over with the dishtowel. The difference is unmistakeable.

I believe the explanation is that water evaporates more quickly from a hot object, and the crockery, since it conducts heat less well, stays hot longer.

Or for another example in the kitchen, imagine how different removing the toast from the toaster would be if bread were made of metal!

10 September 1990

50 watt vs. 100 watt light bulbs

What's the difference between a 50 and a 100 watt bulb? The first part of the solution is obvious. Assume that a light bulb is a filament of length L and radius r, and let V, R, and P denote voltage, resistance, and power. Then $P = V^2/R$ and $R \propto L/r^2$. Since V is a constant, this gives us

$$P \propto r^2/L. \tag{1}$$

So maybe the 100 watt filament is half as long as the 50 watt filament. Or is it $\sqrt{2}$ times as thick? Or some combination of the two?

Obviously there's a piece of physics still missing from the derivation. To give light at the same wavelengths, the two filaments must have the same surface temperature. But this means that P must be proportional to surface area. So the second equation is

$$P \propto rL. \tag{2}$$

Solving (1) and (2) leads to the delightful conclusion that within a family of light bulbs, R is proportional to L^2. To go from 50 to 100 watts, you increase the length of the filament by the factor $2^{1/3}$ and the radius by the factor $2^{2/3}$.

14 January 1991

The Sisyphyus function

I have a favourite example of a function that is differentiable with derivative equal to 1 almost everywhere, yet whose value is more or less constant.

Pick a large and well-defined group of people — say the set of employees of IBM. Let $f(t)$ denote their *average age at time t*. Then $f'(t) = 1$ for all t except at those instants when someone is hired or fired or retired, a set of measure zero. Still $f(t)$ never changes much.

15 January 1991

Splines of iron

At the station last night, I noticed that railway trains provide a good example of splines. The track is an arbitrary curve, and the train is a continuous, piecewise-linear approximation defined by certain interpolation conditions.

8 April 1993

Marital Doppler

Yesterday I drove from Ithaca to Troy, 200 miles away. Now Anne is pregnant, with the baby due any day. Thus it was important to telephone home en route, so that I could rush back if she went into labor.

Suppose I telephone home every half hour. Then since each call comes from half an hour further along the route, my potential return times are spaced at *one hour intervals*.

This is nothing but the Doppler effect. If my calls occur with frequency f, think of each of them as emitting a virtual particle, a potential Nick Trefethen driving back home at highway speed. Because of the actual Nick Trefethen's equal velocity in the opposite direction, those virtual particles arrive in Ithaca with frequency $f/2$. One octave lower. Red-shifted.

Similar effects would apply, for example, to the Grande Armée marching from Paris to Moscow.

27 February 1997

Multimedia encounter at Rhodes Hall

Rhodes Hall at Cornell is built in a gentle curve. Last night as I walked along the building, a car was approaching, and I noticed that the sound from it seemed fuller than usual. Might the concave face of the building be focusing the sound at me, I wondered?

I looked up and saw the car's headlights reflected in the windows of the building. Not just in one window. The reflection came from this window, and that one, and the one after that — half a dozen in all. The curve of the building face made the angles just right for all those windows to reflect the light at me at once.

This all happened in a second or two, and the experience was exquisite.

11 January 2002

Pitcairn Island: a paradise for Muslims

A Muslim, we know, must face Mecca when he prays. But what if he lives on the antipodal point of earth from Mecca? The lucky fellow can face any way he likes!

From my atlas it seems that Mecca's antipode is the South Pacific atoll known as Tematagi. But close enough and more memorable, just a few hundred miles away, is Pitcairn Island. From now on I will think of Pitcairn Island as Islam's special point of isotropy; though confusingly, it was settled by a Christian.

6 July 2002

Why did we buy an oboe but rent a cello?

The oboe was for Emma, the cello was for Jacob, and the answer goes straight to the laws of physics. If you shrink a cello but at the same time adjust the tension or density of its strings, you can play the same notes as before. So there are small-sized stringed instruments for kids; who outgrow them; so it makes sense to rent.

But you can't shrink an oboe. The compressibility and density of the air can't be adjusted, so the pitch would have to go up if you shrank it. Therefore kids play the same instruments as grownups, and it makes sense to buy.

22 August 2004

Why more women than men have AIDS

Anne Morris is a specialist in HIV and AIDS, and she showed me one of her PowerPoint presentations. The message is remarkable: whatever may have been true in the past, today AIDS is a disease of women as much as men and of heterosexuals as much as homosexuals. Indeed, in Africa it afflicts more women than men.

One might think that this is because "men sleep around more than women"; but that is not the explanation and indeed it is mathematically dubious since each time a new couple has sex, one man has a new partner and also one woman has a new partner. No, the asymmetry is in infectivity, not behaviour. Think of an extreme. Suppose men could infect women but not the reverse. Then in a world with plenty of partnering and all heterosexuals, you could have just a few infected men and arbitrarily many infected women.

Taking one step towards reality brings in a square root. Imagine a heterosexual population with a small fraction of infected men and women. Suppose that in each encounter, an infected man is nine times as likely to pass on the disease as an infected woman. Then if three times as many women as men are infected, we have a steady state: a random encounter, on average, infects a new man one-third as often as a new woman.

30 September 2008

The mathematics of spinsterhood

Kate recently reviewed a book called *Singled Out: How Two Million Women Survived Without Men After the First World War*. What a magnificent regiment these "surplus women" made! It seemed everyone had a maiden aunt in those days, and some of these women became the backbone of their generation.

The statistics are uncertain, but it seems the war killed about a million British men; so roughly speaking, a million British women spent their lives single as a result. And the book is filled with their stories.

Now you might easily fall into the mistake of thinking that these were woman who lost a fiance or a husband. But of course the truth must be that there were many too who never found one, who stayed single because in the shrunken pool of post-war men, they had fewer dates and fewer romances than they would have had otherwise. It's easy to model this situation, as it were, mathematically — making a dozen simplifications as mathematical modelers always do. For each of the half-million soldiers (say) who died married or engaged, some woman lost a partner. For each of the other half-million who died unattached, some woman never found a partner.

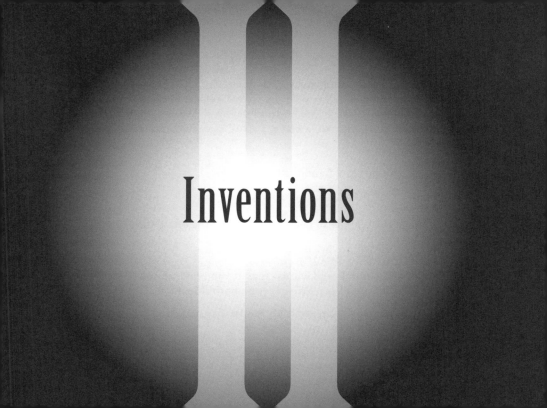

1 January 1986

Improved rules of Scrabble

Chris Morris, Nathaniel Foote, and I have devised a new set of scoring rules for Scrabble, aimed at making the game more verbal and less mechanical. One ignores all double and triple bonus squares, and also the special 50-point bonus for playing seven tiles at one turn. Instead, a player's score at each turn is the sum of three quantities:

(1) The usual Scrabble score (ignoring bonus squares)

PLUS

(2) The *square* of the number of new tiles placed on the board this turn

PLUS

(3) The *square* of the number of words created, modified, or crossed this turn. To "cross" a word means to place new tiles on both sides of it.

The new rules seem to work pretty well. My mother has suggested the name "Scrabble Squared".

27 May 1986

Fibonacci currency

Dollars should be printed at a Fibonacci rate.
I'll give you this thirteen for a fiver and an eight!

And if the thirteen won't suffice, you need a little more,
Just add it to a twenty-one and buy a thirty-four!

29 March 1987

Dog toilets

Several tons of dog excrement are produced each day in a city like New York — deposited outdoors, then laboriously scooped up by their masters for proper disposal. But dogs are neither stupid nor clumsy. Can't some kind of toilet be invented that they could be trained to use?

I predict that dog toilets will appear on the market before too long. Keep your eye on Bloomingdale's.

30 March 1987

Hexagonal highways

The U.S. has a rectilinear grid of interstate highways — even numbers running roughly east-west, odd numbers north-south. In principle, this scheme implies that some journeys will be $\sqrt{2}$ times longer than the distance between the endpoints. Fortunately, the grid deviates enough from the rectangular pattern that it practice it is rare to pay such a large mileage penalty on any long trip.

Still, I wonder whether other geometries would work. What about a hexagonal grid, with three preferred axes? The theoretical maximum mileage penalty would drop from 41% to 15%. Has this been tried anywhere?

25 April 1987

The *N*-clock

Every conference room and lecture hall should be fitted with an *N*-clock. At the beginning of the meeting you set the elapsed time to zero and punch in *N*, the number of people present. Then the clock starts running — at *N* times the usual rate. It's marvellous how much needless chatter is avoided when everyone can see that minute hand swinging round the dial.

30 December 1990

Changes brewing

I predict a revolution ahead in the way we make coffee. The current most popular method is filtering, which delivers a tasty result but is labor-intensive and messy. So restaurants brew a whole pot at a time, and unless they're unusually careful or have plenty of volume, the pot burns on the hotplate and soon becomes unpalatable. Individuals at home produce the same result, or save trouble by using instant coffee that's unpalatable to begin with.

The result: here in the USA in 1990, both in restaurants and at private dinner parties, the quality of the coffee is the least predictable part of any meal.

Coffee is too central a feature of American life for this unsatisfactory situation to persist. In the tradition of food processors and microwave ovens, I foresee the widespread introduction into our kitchens of a newly conceived, touch-of-a-button, cup-at-a-time coffee machine. Or maybe all that's required is a breakthrough in the formulation of instant coffee. Whatever the technology turns out to be, somebody will make millions from solving the coffee problem.

5 May 1996

Disposal, Inc.

I see an opportunity for a new company. This is not a local operation but a nationwide one, with an office in every town.

Let's call the company Disposal, Inc. Let's give it a motto: "We'll take anything off your hands — don't let it hang around!"

Here is what Disposal does. It takes away anything you don't want anymore, from a rusty broken dishwasher to a working color television, from dirty motor oil to antique books. And here's the trick: sometimes you pay Disposal for the service, sometimes Disposal pays you, and sometimes no money changes hands at all. Disposal picks things up at your house, if you need that, and it covers the whole range — taking some things to the dump, others to the Salvation Army, some to a local antique store, some to a high-class auction house in New York.

Disposal names the price, and you take it or leave it. It's not top dollar, but you trust its reputation of being reasonable both financially and environmentally.

Millions of households have things they would dispose if only it were effortless to do it right. Disposal will make it effortless.

27 August 2000

Who died today? — the TV show

There have been dramatic disasters lately. A supersonic Concorde plane crashed for the first time, killing 143; now all Concordes are grounded. The Russian Kursk submarine sank and we had a week of drama until it was revealed that all 118 crew were dead. These events swamped the world's television and radio news.

Meanwhile — a familiar observation — the deaths from other sources pile up in far greater numbers without much press attention. Take car accidents. I don't know the figure, but worldwide, 1000 people or more must be killed by cars each day.

Here's an idea for a radio or television program — has it been tried? Pick a day in advance, get your worldwide news correspondents primed to collect the needed information, and then broadcast a special report the day after, one by one, on each of that day's thousand. Redheaded 18-year-old Sally Smith died in Topeka when a drunk driver hit her; she was going to start at the U. of Kansas in September. Farmer George Thomas must have fallen asleep at the wheel five miles outside Peterborough; he and his wife Ellen were killed instantly when their car hit a tree. And on and on for an hour, with just enough human detail to make it real.

15 October 2005

The Google Game

We spend our lives on Google looking for this and that. It's interesting to consider the inverse problem, what might one search for to find oneself? Here are the precise rules of the *Google Game*: What words can you find, or pairs of words, such that when you type them into Google, the first hit is a page by or about yourself? A pair of words is the limit; you're not allowed to type in three or more.

Here some Google roots I've found for myself.

pseudospectra	SCPACK
Kreiss matrix	Talbot contour
spectral methods	without eigenvalues
numerical professor	stiff PDE
science maxims	SIAM challenge

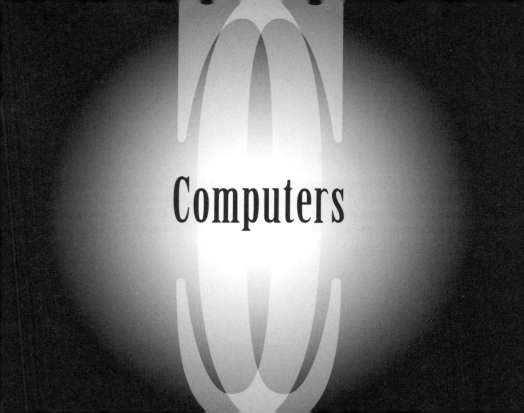

6 April 1979

Computer doom

Will computers and their descendents bring a horrible future down upon our heads? I set this down on the basis of a number of conversations: most people in the field today believe that information technology does not pose a significant threat to our future well-being. How can they be so confident? I am more apprehensive than that.

1 October 1987

How to do envelopes in 1987

Computers are taking over in offices; Selectric typewriters are giving way to laser printers. But the advance is uneven, and has its comical side. In particular, nobody in this modern era has figured out how to do envelopes.

And so it was that I received a letter the other day from Donald Knuth. Knuth is probably the most famous computer scientist in the world, and a professor at one of the two or three most influential (and richest) computer science departments: Stanford University. My address on this letter had been laser-printed on a sheet of paper, cut to size with a pair of scissors, and Scotch-taped to the envelope.

3 April 1988

Is it OK to treat a computer as a slave?

We leave more and more tasks to machines. Today they answer the phone and record TV programs. Tomorrow we will call from work to ask them to brew a pot of coffee.

Once, wives or servants performed such tasks. If machines are to do them in the future, one naturally wonders, what is the difference between people and machines that makes this division of labor sensible?

The answer must be that in some sense machines are cheaper, but I think this is true in part for a strange reason: they can be treated as slaves rather than equals. In theory a human slave might be maintained for $1000 per annum, outperforming electronic competition at many tasks. But to maintain a person at a humane level of comfort and dignity costs ten times as much or more. In contrast, nobody demands that we worry about the comfort and dignity of a computer.

Can the difference persist as machines grow more intelligent? The day may come when it is considered inhumane to force advanced computers to do routine work for 24 hours a day. Fortunately, more primitive devices may still be on the market whose dignity can be ignored. Perhaps one day it will be seen as an advantage of machines that they can be manufactured to lack certain capabilities, as well as possess them.

9 August 1988

Paradox of computer science

At the heart of computer science lies a paradox: as computers become more powerful, the efficiency of algorithms becomes *more* important, not less.

This principle seems cockeyed at first, but its explanation is not difficult. To be sure, a faster machine will make some old problems so easy that any algorithm can handle them. But there are always bigger problems waiting to be solved. As the speed of machines increases, the efficient algorithms diverge dramatically from the inefficient ones in the *size* of problem they can handle. Thus as in so many other areas, fuzzy distinctions become sharp ones — indeed, a science is born — in the limit as something approaches infinity. Theoretical computer science is a young field but surprisingly mature, and the reason is that the power of our machines is already quite near infinity.

This paradox deserves a name; has it got one?

3 March 1989

The loss of determinism in computing

Present-day digital computers are essentially deterministic. Even the rounding errors, though random in a sense, are utterly reproducible in most cases, thanks to extraordinary care taken to check for errors at all levels in the hardware. For example, one sees this dramatically in certain fluids calculations: no matter how unstable the flow may be physically, if the initial and boundary conditions are symmetric and so is the computational method, no asymmetry will ever appear in the calculation.

Will computers remain deterministic as the decades pass? I suspect the answer may be no. Of course we can have determinism if we want it enough, but I doubt we do, for it tends to restrict us to primitive methods. Think of a typewriter vs. TeX, or even old photocopying machines vs. modern intelligent ones. The smarter the system gets, the harder it becomes to predict its behavior, and the less often we take the trouble to do so. In thirty years our computers and indeed all our tools may be so flexible in the ways they help us — adaptively granting resources as available to match our demands as they arise — that we'll be unable to get exactly reproducible results except by extraordinary efforts. The sacrifice of determinism will be the symbol of progress to another level of operation.

15 July 1995

Billions of underutilized brains

Tens of millions of computers around the world — the figure will soon grow to billions — are largely unutilized. Vast periods of idleness separate brief flashes of computation. At the end of their lives, most computers have probably done less than 10% of the thinking they might have done.

One's first reaction is to be shocked by such waste. What a scandal! A little reflection, however, reminds us that our dishwashers and lawnmowers and automobiles are equally underutilized, without anybody getting upset about it. It is hard to come up with a compelling reason why brain machines must be subject to different rules from muscle machines.

But here, an analogy hits. There are five billion human brains on the planet, and most of them are as underutilized as most computers. Most people, even talented ones, lead small lives of narrow ambitions, their higher faculties more or less idle. This seems shocking — but is it, really? Must we feel obliged to utilize fully a merely good machine, when outstanding ones are available elsewhere? — when it doesn't happen to be loaded with the best software? — when, after all, it's an easy matter to make more?

3 March 1997

Hamming updated
The purpose of computing is insight, not pictures.

21 March 1997

Parallel brains, parallel computers

Nobody really knows how to program parallel computers. Nobody really knows how the brain works. In the next century, related revolutionary developments will occur in both areas.

18 January 2002

President Jefferson at breakfast

One day in January 1802, President Jefferson finds a note at the breakfast table.

Dear Mr. President,

At lunchtime two hundred years from today, a professor in Oxford will suggest to a student across the hall and a colleague in Texas that they send a message to a company president in California proposing that the four of them fly through the air on the first Monday in June, at a total cost of a few days' income, to get together for an afternoon in Boston. The colleague in Texas will respond a minute later, suggesting that the three of them first create some color images for the company president to look at. The student will spend an hour producing the images, in the course of which he will perform about 100,000,000,000 additions, subtractions, multiplications and divisions to 16-digit accuracy without any mistakes. He and the professor will then send their message and pictures to the man in California. As the sun sets, the professor will expect it is likely he will have a conversation about them with the man in California before bedtime. He will hope it does not interrupt his evening viewing of talking, moving pictures of the day's news events from Afghanistan, central Africa, Argentina, China and Ireland.

Life and DNA

6 September 1971

Man's invisibility on most spatial scales

We tend to picture mankind as something "unnatural" — something that somehow goes beyond ordinary deterministic or random laws of nature. This point of view would imply that, when we look at the universe, man will stand out as something special.

On a large scale, that is not (at least now) true. Looking at the whole universe, man is a practically infinitesimal speck.

On a small scale, it is even more untrue. The individual atoms have not changed their behavior since *Homo sapiens* appeared.

Only in the in-between scale is man's presence detectable: in the macroscopic collections of matter.

(Age 16)

16 June 1973

Life as molecular amplifier

It is probable that a shifting of a relatively small number of molecules in Richard Nixon's head could cause an H-bomb to be dropped on Moscow; in this sense presidents, and humans in general, seem to be particularly powerful amplifiers of microscopic events. Such amplification is presumably a characteristic of all life. If I could only get my hands firmly on the concept, that fact might even be a reasonable definition of life.

(Age 17)

15 July 1976

Consciousness and moral obligation

Just as moral obligation is undefinable, it has long seemed to me that consciousness, that is self-awareness, is undefinable also. This apparent parallel has nagged at me painfully, for I "believe in" consciousness, but not in moral obligation.

28 July 1985

Death as ecological disaster

From the point of view of an individual cell in my body, the universe consists of itself and many similar fellows, closely packed together, and bathed in the nutrients needed for survival. The environment is benign, well adjusted to support life.

When I die, it will seem to this cell that the life-giving ingredients of the oceans and atmosphere have been inexplicably replaced by poisons. The cell will die, victim of an ecological disaster.

23 August 1995

Jellyfish more miraculous than dogs

Recently we saw a special exhibit of jellyfish at the aquarium. What was special was the focus on this single group of organisms and the beautiful illumination. Well-placed lights showed these creatures pulsating, beating, moving constantly, expelling water behind them and propelling themselves downward.

Such simple, transparent bodies, so elementary yet so alive! One could only think, what a miracle life is!

When we look at a dog or a horse or a person, we are less struck by the miracle of life. The gulf between jellyfish and inanimate substance is grand enough, and unfamiliar enough, to be awe-inspiring. The gulf between higher animals and inanimate substance is so vast, yet so familiar, as to be hardly conceivable.

13 March 1997

Solving the problem of consciousness

Eventually mankind will solve the problem of consciousness by deciding that we are not conscious after all, nor ever were.

26 April 1997

Computer codes and genetic codes

Computer codes are better written than genetic ones, since there's a programmer in the loop, but as they get bigger, this distinction is fading.

30 April 1997

Life is one-dimensional

Living things are 3D objects, yet they are constructed by folding up 1D pieces. This astonishing method of construction is what makes repair, reproduction, and natural selection practicable. If an infinite intelligence designed an organism from scratch, by contrast, presumably it would use 3D building blocks.

18 June 1998

A trivia question you'll never be asked

Q: Along with the usual hydrogen, carbon, nitrogen, and oxygen, there is a fifth atomic element that appears in two of the twenty amino acids from which all life on earth is made. What is this element?

A: Sulfur.

This little question epitomizes the gap between the two cultures. If you play Trivial Pursuit or watch Jeopardy for a few years, you may be asked about Michael Jordan's middle name, or the most popular movie of the 1930s, or the author of *Treasure Island*. But I promise you, you will never need to know that sulfur is the fifth element in amino acids.

18 February 2001

Ancient Trefethen DNA

Two curiously related events happened this week.

One was that Celera (in *Science*) and the Human Genome Project (in *Nature*) published their articles announcing the sequencing of the human genome. And there's a surprise: a human has only 30,000–40,000 genes.

Another was that I got an email message from a certain Grace Griffin in Haverhill, Massachusetts telling me that my great-great-great-great-grandfather Henry Trefethen and her great-great-great-great-great-great-grandfather John Trefethen were brothers.

This set me thinking. How much Trefethen DNA have Ms. Griffin and I inherited in common? I believe 1/64 of my DNA comes from Henry, and 1/256 of hers comes from John. Since Tom and Henry had half their DNA in common, this implies that she and I share about $2^{-15} = 1/32768$ of our DNA.

So my new-found cousin and I have about one Trefethen gene in common.

4 July 2005

Google PageRank and the brain

Google has taken over the world, and it is based on a beautiful idea. Suppose one surfs the web, following links at random. Among the billions of all web pages, what fraction of time will one spend at a particular page k? That fraction is its *PageRank*, p_k. Some nodes are more important than others, being often visited, and some links are more important, being often travelled.

No doubt I am not the first to wonder about analogies with the brain. There we have millions of neurons, each connected to a variable number of other neurons near and far. It seems that links that are often travelled become reinforced somehow. Is there a Markov chain lurking here, approximately speaking, and does each node in the brain have a PageRank?

15 June 2006

Both feet the same size

Our friend D. was over for dinner, and he is unusual: his right and left sides are not the same. In particular, his left eye is brown and his right eye is green. Perhaps, we wondered, he is some kind of chimera, with different genes inherited from his parents expressed on the two sides of his body?

Many people have one foot bigger than the other. D. told us significantly, however, that his two feet are just the same. Further evidence of asymmetry!

11 April 2008

Icelanders

Iceland has one of the purest, most inbred populations in the world. From the 13th to the 20th century, not many people flowed in and out. The genetics company deCODE was founded to take advantage of this excellent dataset to track genetic traits. In Iceland we know for generations back who is whose son and who is whose daughter.

Some of deCODE's discoveries have been published this week in *Nature* in an article called "Genetics of gene expression and its effect on disease." Here are the authors:

V. Emilsson	G. Thorleifsson	B. Zhang	A. S. Leonardson
F. Zink	J. Zhu	S. Carlson	A. Helgason
G. B. Walters	S. Gunnarsdottir	M. Mouy	V. Steinthorsdottir
G. H. Eiriksdottir	G. Bjornsdottir	I. Reynisdottir	D. Gudbjartsson
A. Helgadottir	As. Jonasdottir	Ad. Jonasdottir	U. Styrkarsdottir
S. Gretarsdottir	K. P. Magnusson	H. Stefansson	R. Fossdal
K. Kristjansson	H. G. Gislason	T. Stefansson	B. G. Leifsson
U. Thorsteinsdottir	J. R. Lamb	J. R. Gulcher	M. L. Reitman
A. Kong	E. E. Schadt	K. Stefansson	

Hearts Minds and Bodies

4 August 1970

Four wishes

Some of my material wishes have been

(1) To be able to fly
(2) To be able to read minds
(3) To be able to become invisible
(4) To be able to dream whatever I want.

(Age 14)

12 March 1972

Why can't we smell water?

Why are we unable to smell water? Moisture affects many chemical reactions, so it should be something easy to be sensitized against, and it is important to survival; thus I would have expected sensing of water to have evolved. Why not?

(Age 16)

29 March 1981

Oceans

Oceans, mountains, or stars irresistibly put one in a profound, philosophical state of mind. (This is not encouraging evidence for the validity of our philosophies.) Oceans are best, for they have not only immensity but a grave, inhuman life. An ocean has a powerful effect on me.

23 April 1985

Why we sleep

The question of why we sleep is not yet answered. But the view is widespread that sleep is needed to permit the brain somehow to sort out or recover from the experiences of the day.

Something of this kind may be the proximate explanation, but I suspect the deeper reason is the existence of nighttime. We are diurnal in feeding habits. Given this, it is plausible that evolution would have contrived nocturnal sleeping to conserve energy. Nobody proposes that bears need to hibernate in order to sort out the experiences of the summer.

5 June 1986

Two worlds designed to suit us

We live in two worlds. There is the natural world of earth, air, sunlight, rivers, strawberries. And there is the man-made world of cars, streets, money, post offices, Coca Cola.

Both worlds are perfectly fitted to our needs. If there were no fresh water on the land, we would die of thirst! — but there is. If there were no gas stations, my car would be useless! — but they are everywhere. We are so accustomed to this favorable state of affairs that the slightest truly random element in our environment is easily interpreted as an act of hostility, a breach of trust.

What strikes me is the 180° difference in *why* these two worlds are so well fitted to us. First came nature, and *we* adapted to *it*. Then came culture, and *it* adapted to *us*. In one case we are the product, in the other case the designers — and yet both fits are so good that we rarely notice the distinction.

15 August 1989

Are short people fat?

In keeping with the stereotype, are short people fatter on average than tall ones? This is an easy empirical question, and its answer is surely known.

Assuming the answer is yes, here is a hypothesis to explain it. Our meals are designed around an average eating pattern, and the size of the portions does not automatically adjust to the size of the eater. Yet to retain a similar shape, two people differing in height by 10% should have food intakes differing by 33%, thanks to the power of cubes. Thus, far from encouraging people of different heights to maintain similar shapes, our society actually encourages people of all heights to weigh the same number of pounds.

26 May 1995

Addiction

When the body depends on a substance — when the spirit craves it — when deprivation renders all other goals subservient to the obsessive need to get more — when continued deprivation leads to illness and violent crime — this we call an addiction.

The habit of eating food is an addiction of the most definite kind. Prolonged deprivation may lead to death; and as to the earlier stages, the effects on the mental state are legendary. Look what happened to Jean Valjean.

Society, however, does not view this particular addiction as depraved. In fact, people routinely come together to perform the act of eating in groups, inserting food in their mouths and chewing and swallowing it communally. At the close of such occasions, the participants may go their ways without the least shame or furtiveness, only to come together to repeat the event when the craving rises up in them again a few hours later.

The reason that eating is tolerated by society, of course, is that the addiction is universal and no cure is known. Certainly the costs are tremendous. A shockingly high proportion of our industry is devoted to the preparation of plant and animal parts to support this habit, and the loss of productivity associated with addicts' constant interruption of their lives for eating is incalculable.

6 September 2002

Extinction of species

This week I finished Lomborg's *The Skeptical Environmentalist* and also E. O. Wilson's *The Future of Life*. Quite a pair!

The central fact about extinctions is not that the earth is becoming uninhabitable, but that it is becoming biologically globalized. If we are still here in a thousand years, we'll have largely the same flora and fauna all over the place, at least along each latitude, and I think Wilson is right that half of species currently on earth may be extinct and that most areas of wilderness will be gone. In a word, the earth may be everywhere like England today: clean, healthy, and beautiful, but nothing like its original wild self.

This is very sad. It's painful; it's moving. However, I cannot see that it is more painful than other aspects of our loss of heritage. A century ago there were many cultures on earth; increasingly we are heading towards one global culture. The nations of Europe are being homogenized as we watch. A century ago thousands of languages were spoken on earth; now the number is hundreds, and falling. No DNA is involved, but in any practical sense, these changes are irreversible too.

I do not think the extinction of species will prove to be among the ten biggest challenges and terrors in store for us in the next millennium.

29 January 2004

Brownian motion by a team of ants

I looked at the pavement and saw the most remarkable thing: a piece of cracker, maybe two centimeters across, moving steadily towards the wall. I looked closer and saw a few dozen ants all around it, pushing it along with their collective strength. Amazing! I'd never seen ants collaborate like this on a big engineering project.

Then the piece of cracker changed direction and moved another way entirely. And then again, yet another direction. And I realized that it wasn't going anywhere purposeful. The ants weren't collaborating at all, just munching away and moving the cracker by accident. Its motion was random, the net result at each moment of the arbitrary forces exerted by a hundred little hind legs. A formic rugby scrum.

Once I saw what was going on it was wonderful to step back, blur my eyes, and watch the cracker swim about. This was Brownian motion, except with forces supplied not by molecular collisions but by earnest little ants.

31 December 2007

Birdbox

For Christmas Kate's Uncle John made her a birdbox, or as I grew up calling such things, a birdhouse. It's a beautiful solid structure that for some lucky blue tit will provide a life like leasing a suite in the Grand Plaza. There's a little ledge with a round hole just the right size for the bird to get in and out, with walls thick enough to keep out any predator. The lucky discoverer will find space and security to raise a happy family.

I find myself thinking, how will the bird who comes across this perfect structure explain its existence?

But of course, birds don't think about such things. They don't try to explain the existence of a perfect birdbox. They just live their lives and that's that.

And then I find myself thinking, what is in fact the true explanation of the existence of this birdbox? Why did Uncle John build it? Why did Kate install it in her garden? The more you think about it the more you realize that the answers are bound up in extraordinary complexities of our emotions about caring. In the end the existence of Kate's birdbox is too hard for humans to explain, let alone birds.

28 April 2008

Shadowlands and Ratatouille

A while ago in London, Kate and I saw "Shadowlands" one afternoon and "Ratatouille" the same evening.

Shadowlands is as serious and sad as a play can be. C.S. Lewis muses deeply on the profoundest questions of life and God and falls in love with young Joy, only to have her agonizingly taken away by cancer. The acting was powerful and we were moved. In fact we were on the edge of tears.

Ratatouille is an animated Disney movie for the whole family about a rat in Paris with the olfactory genius to be a great chef. To get there he has to befriend a clumsy animated young kitchen assistant and his snooty animated female supervisor, along with the wicked animated restaurant owner and the terrifying animated food critic Jacques Gusteau. He does it all and the restaurant becomes celebrated across Paris.

Ratatouille was as moving as Shadowlands. We were brought to tears again. We left the cinema with wet eyes.

Human emotions are the notes of a violin string. If you hear them singing beautifully, don't assume anything beautiful must be the cause. It is their nature to sing.

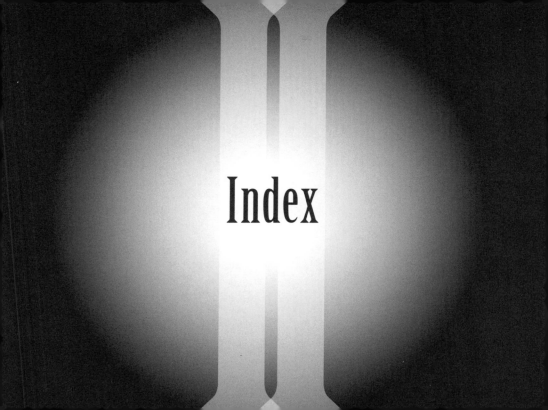

accidents   17, 31, 220
addiction   354
adulthood   x, 13, 17, 23, 85, 246
aging   22, 24, 29, 31
AIDS   80, 223, 311
alcohol   179
algorithms   x, 274, 282, 283, 327
ambition   2, 3, 4, 7, 8, 208, 329
amino acids   341, 342
analogies   146, 148, 201-212, 264, 294
antonymy   203
ants   356
appetites   23
asteroids   249
atheism   232, 233
Australia   x, 48, 70, 152, 171, 267, 287

babies   12, 36, 38, 214, 307
baseball   97, 109, 171, 290
Bach, Johann Sebastian   90, 107, 112
Balliol   91, 92, 95, 102, 176
Beatles   144
beauty   34, 41, 42, 241

Beethoven, Ludwig van   112
behaviorism   105
Berlin, Isaiah   112
bicycles   17, 46, 98, 128, 220
birds   167, 230, 357
birthday   4
bits of information   89
Bjørstad, Petter   128
Blair, Tony   92
books   6, 10, 18, 24, 95, 126, 161, 163, 165, 170, 178
Borwein and Borwein   27
brain drain   207
brains   170, 172, 182, 237, 252, 253, 264, 292, 329, 331, 344, 351
bribery   67
British Rail   53, 95, 224
Brownian motion   356
Bryson, Art   189
Burrage, Matt   212
Bush, George W.   237, 249
Butler, Samuel   202

Cambridge   84, 90, 100

capitalism 72, 195
car accidents, 185, 215, 321
carpe diem 30
cats 40
cello 16, 302, 310
chaos 10, 156, 220, 284
Chapman, Jon 221, 298
chewing 18, 354
Chinese whispers 212
Chomsky, Noam 107
Christianity 229, 233, 237, 243
Christ, Jesus 111, 237, 238
Christmas 238, 357
Churchill, Winston 3
chutzpah 52
cigarettes 77, 215
class 57, 65, 91
clichés 146
Clinton, Bill 92, 218
clocks 180, 318
clothing 65
coffee 152, 184, 319, 326
combinatorics 196

comma splices 95
commute 120, 245
computer programs 276, 288, 331, 340
computer science 280, 294, 325, 327
consciousness 179, 336, 339
conspiracy 48, 95, 107
continuous assessment 84
contraception 57, 82
Copernican Principle 9, 200
Cornell University x, 308
cricket 97
cuddling 35, 38

data 214, 258, 346
death 22, 26, 57, 80, 215, 248, 321, 337
deCODE 346
democracy 3
determinism 252, 253, 328, 334
desire 34, 40, 241
Devil 234
Diamond, Jared 18
Dickens, Charles 90
discrete and continuous 280, 296

dish washing   181, 303
Disposal, Inc.   320
dissonance   203
DNA   343, 355
dogs   149, 179, 316, 338
Dole, Bob   218
doom   5, 324
Doppler effect   307
drives   36, 39, 47, 56, 91, 96
driving   19, 71, 121, 185, 215, 237, 307, 317, 321
drugs   179, 223, 248, 354
dual of age   28
Dylan, Bob   141, 144, 167
dynamics   10, 41, 284
Dyson, Freeman   160

ecological disaster   248, 337
economists   238, 285
The Economist   126
Edison, Thomas   113
editing   24, 48, 144, 151, 159, 163, 193
Eiffel Tower   113
eigenvalues   142, 211, 212, 322

Einstein, Albert   3, 90, 257, 265
electrocution   94
Eliot, T. S.   167
emotions   13, 26, 60, 96, 357, 358
England   x, 8, 15, 90, 91, 95, 98, 101, 207, 303, 355
esthetics   241
ethics   58, 168, 241, 245
evaporation   207, 303
evil   107, 240, 241
evolution   12, 36, 57, 185, 188, 351
extinction   76, 355
extraterrestrial life   265, 268

fame   10, 111, 291
Fibonacci series   315
final causes   195
fire   254
First Cause argument   242
fitness   206
floating point numbers   296
fluid mechanics   6, 266
Foote, Nathaniel   64, 78, 314
fossils   106

Fox, Geoffrey 174
fractals 256
fracture mechanics 288
France 8, 15
freedom 125, 215
French 4, 108
friends 6, 50, 64, 70, 161

Gander, Walter 151
gargoyles 230
Gates, Bill 108
Gauguin, Paul 42
geek 54
genetic code 246, 292, 340, 343, 346
genius 285
German 4, 86, 147, 151, 182
Gilbert, John 258
girlfriends 50, 64
Glass Bead Game ix
God 58, 60, 167, 188, 225–238, 246
Google 12, 322, 344
Gott, J. R. 9, 200
Gramlich, Carolyn 229, 262

Griffin, Jasper 95
grooks 110, 119, 234
guns 18, 77, 223, 237
Guns, Germs and Steel 18

haloes 228
Hamming, Richard 330
Handel, George Friedrich 107
harmony 142, 294
Harvard University x, 108
heaven 243
hedgehogs 112
Hein, Piet 110
hibernation 351
highways 19, 317
hiking 72
Hiroshima 74
homesickness 48
honors 7, 69
Horace 167
horses 121, 154, 338
Hoyle, Fred 7
hubris 248, 257

Hume, David 200
humor 203
hunger 39

ice cream 32
Iceland 20, 346
ideas 51, 85, 104, 112, 159, 160, 162, 165, 189, 191, 192, 193, 196, 282
ill-conditioning 300
immortality 62, 227, 229
importance 196
inconsistency 168, 244, 250
index cards ix, 12, 85, 160, 167, 191, 194, 196
induction 200
infinity 327
instability 195
intelligence 36, 56, 179, 341

Japan 98, 102
jargon 189
Jefferson, Thomas 332
jellyfish 338
Jetsons 68

kickstands 98
Knuth, Don 114, 325
Kublai Khan 102
Kuhn, Thomas 256

Langstaff, Lee 230
languages 47, 147, 180, 197, 355
lasers 212, 325
Latin 151
laws of physics 252, 257, 296, 310
lawyers 67, 87, 208
Lax, Peter 114
learning 86, 182, 189, 254, 262
light bulbs 94, 122, 304
logic 54, 182
Lottes, James 298
luck 66, 111, 294, 357

madness 46
marijuana 179
Markov chain 344
marriage 4, 116, 233, 312
Marx, Karl 285

Mathematical Engineering 278
Matlab 124
McLoughlin, Kate 54, 92, 167, 312, 357, 358
Mecca 309
meltdown 148
memory 24, 30, 169–176, 276
men and women 15, 25, 36, 236, 311
microwave ovens 71, 209, 319
middle age 2, 22, 128
Mill, John Stuart 104, 192
MIT x, 48, 132, 135
mnemonics 176
mobile phones 71, 88
molecules 199, 252, 295, 296, 335, 341
money 72, 208, 238, 276, 320, 315, 352
moral imperative 242, 243, 336
morality 202, 242, 244, 246
Morris, Anne 311
Morris, Chris 314
movies 13, 58, 144, 170, 358
Mozart, Wolfgang Amadeus 3, 90, 106
musical instruments 142, 211, 310

naked ladies 40, 42
Napoleon 108
natural selection 56, 66, 188, 195, 206, 231, 341
nature vs. nurture 204
N-clock 318
needles 223
Newsweek 85, 244
Newton, Isaac 90, 257
New York University x
Nixon, Richard 335
nonlinearity 284
nonsense 203
novels 60, 74, 161
nuclear power 81, 148
nuclear war 73–83, 215, 248
numerical analysis x, 274, 278, 282, 283

obscurity 167, 282
Ockendon, Hilary 175
of course 166
ontological argument 234
optical illusions 32, 198, 228, 263, 270
optimality 10, 126, 127, 163

orthogonality 92, 163
osmosis 148
oubliette 123
Oxford 16, 52, 88, 90, 94, 100
Oxford University x, 9, 88–90, 92, 102, 134, 332

PageRank 344
paradox 41, 117, 206, 327
parallel computing 122, 331
Pascal's wager 76, 229, 247
Paul, David 64, 171
personality ix, 6
Phillips Exeter Academy x
philosophy ix, 131, 190, 199, 232, 241, 254, 350
piano 4, 86, 126, 141, 142
Picasso, Pablo 90
Pitcairn Island 309
pleasure 3, 30, 35, 161, 210, 229
politics 79, 92, 133
population 222, 290–292
prejudice 184
private schooling 91

prizes 69
probability 75–77, 81, 89, 94, 120, 200, 229, 231, 268
pronunciation 96, 99, 147, 180
punctuation 95
pure and applied 143, 279, 286

quantum mechanics 58, 257

radio 88, 96, 132, 140, 144, 172, 237, 321
randomness 60, 111, 120, 141, 185, 203, 207, 214, 262, 291, 301, 328, 344, 352, 356
reason 92, 192, 199, 244, 264
reflection 301, 308
relationships 49, 50
remainder 28
resonance 162, 165, 211
Riemann hypothesis 281
rigor 199, 284
risk 71, 75, 77, 78, 81, 215
rivers 208, 250, 301, 352
rock and roll 143, 172
rounding errors 274, 296, 328

Rowling, J. K.   285
Royal Society   7
Russell, Bertrand   107, 227

sailboats   259
sailors   14
Salieri, Antonio   106
Schell, Jonathan   76
science   57, 192, 196, 231, 232, 251–260, 284
Scrabble   314
SDI   82
seatbelts   77, 215
secretaries   235
semantics   159, 206
sensitive dependence   253
Servetnick, Marc   72
sex   33–42, 58, 248
sex drive   36, 39
Shady Hill School   x, 85
Shakespeare, William   90, 158, 167
shaving   20, 226
Sheehan, Greg   108, 290
Shoeless Joe Jackson   109

shredder   16, 134
sin   246
Skeptical Environmentalist   355
skiing   35
Skinner, B. F.   105
slaves   326
sleep   12, 14, 178, 351
smell   30, 47, 349
soccer   8
Socrates   192
solar system   41
speed of light   265, 297
spelling   99, 147
spinsters   312
splines   306
spoonerisms   175
Stanford University   x, 46, 128, 264, 325
Strang, Gil   114, 176
Strogatz, Steve   viii, 10
Strunk and White   166
suicide   57, 250
sulfur   342
surrealism   203

swimming pools 202
Switzerland 48
symmetry 328, 345

Tao, Terence 287
taxes 217
teeth 18, 173
temperature 53, 199, 203, 304
tenure 4, 135
tests 87
theorems 162, 273, 288
timpani 112, 150
to-do list 118
toilets 316
Tolstoy, Leo 27, 112, 161
tourism 47
translation 151, 156
trees 72
Trefethen, Anne x, 13, 14, 19, 268, 307
Trefethen, Emma v, x, 13–16, 25, 41, 236, 297, 310
Trefethen, Florence 42, 66, 84, 214, 246, 314

Trefethen, Gwyned 19
Trefethen, Jacob v, x, 8, 14, 16, 17, 20, 38, 200, 310
Trefethen, Lloyd M. vii, 6, 26, 183, 229, 259, 267
trigonometry 267, 270, 301
truth 192, 227, 244
tunes 112, 139, 172, 196
typing 121, 124, 136, 182, 235, 325

value 7, 106, 285
violin 142, 358
voting 66, 70, 218

wallahs 153
war 219, 312
wickedness 91, 248, 250
Wilson, E. O. 222, 355
wine 210
women's liberation 216
workaholics 126
World Cup football 8
writing 24, 85, 164